The PRONOUNCING DICTIONARY of PLANT NAMES

American Nurseryman Publishing Company
Chicago, IL

223 West Jackson Blvd. Suite 500
Chicago, IL 60606

ISBN 1-887632-58-1

Library of Congress Cataloging-in-Publication Data

The pronouncing dictionary of plant names.
p. cm.
Rev. ed. of: The new pronouncing dictionary of plant names.
2nd ed. 1964.
ISBN 1-887632-58-1 (alk. paper)
1. Botany–Dictionaries. 2. Plants–Nomenclature. I. American Nurseryman Publishing Company.

QK9.P76 2006
580.1′4–dc22
2005026515
Printed in the United States of America

10 9 8 7 6 5 4 3 2 1

A

Abbreviatus (a-bree-vee-AY-tus). Shortened.
Abelia (uh-BEEL-yuh; uh-BEE-lee-uh). Southern flowering shrubs.
Abeliophyllum (uh-bee-lee-o-FIL-um). Shrubs of olive family.
Abelmoschus (ay-bel-MO-shus). Musk-scented (seeds).
Abies (AY-beez). Fir.
Abortive (uh-BOR-tiv). Imperfectly developed.
Abronia (uh-BRO-nee-uh). Sand verbena, wild lantana.
Abruptus (uh-BRUP-tus). Abrupt.
Abutilon (uh-BYOO-tih-lon). Flowering maple.
Acacia (uh-KAY-shuh). Flowering trees, shrubs; acacia, wattle.
Acaena (uh-SEEN-uh). Nearly evergreen trailing perennials.
Acalypha (ak-uh-LY-fuh; ak-uh-LEE-fuh). Chenille plant.
Acanthocarpa (uh-kan-tho-KAHR-puh). Thorny-fruited.
Acanthofolius (uh-kan-tho-FO-lee-us). Thorny-leaved.
Acanthocereus (uh-kan-tho-SEE-ree-us). Trailing, climbing cactus.
Acanthoclada (uh-kan-tho-KLAY-duh). Thorny-branched.
Acanthopanax (uh-kan-tho-PAY-naks). Hardy Asiatic shrubs, trees.
Acanthophoenix (uh-kan-tho-FEE-niks). Spine areca.
Acanthophyllum (uh-kan-tho-FIL-um). Grayish rockery plants.
Acanthus (uh-KAN-thus). Bear's-breeches.
Acaulescent (ak-aw-LESS-ent). Stemless, or apparently so.
Acaulis (uh-KAW-liss). Stemless.
Acephalus (ay-SEF-uh-lus). Headless.
Acer (AY-ser). Maple.
Aceranthus (a-sur-AN-thus). Maplewort.
Acerbus (uh-SUR-bus). Harsh or sour.
Acerifolia (a-sur-ih-FO-lee-uh). Maple-leaved.
Acerosus (ass-ur-O-sus). Needle-shaped.
Acetosa (a-seh-TOH-suh). Acid-leaved.
Achene (uh-KEEN). Hard, one-seeded fruit.
Achillea (ak-ih-LEE-uh). Hardy perennials; yarrow, sneezewort.

Achimenes (uh-KIM-eh-neez). Showy summer-flowering pot plants.
Achyranthes (ak-ih-RAN-theez). Bedding foliage plants.
Acidanthera (ass-ih-DAN-thur-uh). Summer-blooming, tender corms.
Acineta (ass-ih-NEE-tuh). Tropical American orchid.
Acmophyllus (ak-MOFF-ih-lus). With pointed leaves.
Aconitum (ak-o-NY-tum). Monkshood.
Acorus (AK-o-rus). Herbaceous marsh plants.
Acris (AY-kriss; AK-riss). Acrid, sharp.
Acrocomia (ak-ro-KO-mee-uh). American feather palm; hairy-crowned.
Actaea (ak-TEE-uh). Baneberry, cohosh.
Actinidia (ak-tih-NID-ee-uh). Woody vines.
Aculeata (ah-kyoo-lee-AY-tuh). Prickly.
Acuminata (a-kyoo-mih-NAY-tah). Tapering to a point.
Acutifolius (uh-kyoo-tih-FO-lee-us). With pointed leaves.
Acutissimus (a-kyoo-TISS-ih-mus). Very acute.
Acutus (uh-KYOO-tus). Sharply pointed.
Adansonia (ad-an-SO-nee-uh). Huge trees of Africa, Australia.
Adenanthera (ad-eh-nan-THEE-ruh). Red sandalwood, Barbados pride.
Adenium (uh-DEN-ee-um). Desert rose.
Adenocarpus (uh-den-o-KAHR-pus). With glandular fruit.
Adenophora (ad-eh-NOFF-o-ruh). Ladybells.
Adenophorus (ad-eh-NOFF-o-rus). Gland-bearing.
Adenophyllus (uh-den-o-FIL-us). With sticky, glandular leaves.
Adenostoma (uh-den-o-STO-muh). With glandular calyx lobes.
Adiantifolia (ad-ee-an-tih-FO-lee-uh). Leaves like adiantum.
Adiantum (ad-ee-AN-tum). Maidenhair fern.
Adlumia (ad-LYOO-mee-uh). Allegheny vine, climbing fumitory.
Admirabilis (ad-mih-RAB-ih-liss). Admirable, noteworthy.
Adnatus (ad-NAY-tus). United, joined to.
Adonidia (ad-o-NID-ee-uh). Philippine feather palm.
Adonis (uh-DOH-niss; uh-DON-iss). Pheasant's-eyes.

Adpressa (ad-PRESS-uh). Pressed together.
Adromischus (a-dro-MISS-kus). Thick-stemmed.
Adsurgens (ad-SUR-jenz). Ascending.
Aduncus (a-DUN-kus). Hooked.
Aechmea (eek-MEE-uh). Epiphytic greenhouse plants.
Aegopodium (ee-go-PO-dee-um). Goutweed, bishop's-weed.
Aeneus (EE-nee-us). Bronze-colored.
Aeonium (ee-O-nee-um). Succulent shrubs and herbs.
Aerides (ay-ER-ih-deez). Tropical orchids, all air plants.
Aeschynanthus (ess-kih-NAN-thus). With shameful (red) flowers.
Aesculus (ESS-kyoo-lus). Horse chestnut, buckeye.
Aestivalis (ess-tih-VAY-liss). Pertaining to summer.
Aethionema (ee-thee-o-NEE-muh). Stone cress.
Aethiopica (ee-thee-O-pih-kuh). From Ethiopia or Africa.
Affinis (a-FY-niss). Related.
Afra (AY-fruh). African.
Africana (af-rih-KAY-nuh). African.
Afrocarpus (af-ro-KAHR-pus). Yellowwood.
Agapanthus (ag-uh-PAN-thus). Lily of the Nile, African lily.
Agaric (uh-GAR-ik). A mushroom or like one.
Agaricus (uh-GAR-ih-kus). Common cultivated mushrooms.
Agastache (uh-GAS-tuh-kee, a-guh-STAY-kee, a-guh-STAH-kee). Perennial herbs with many (flower) spikes.
Agathaea (ag-uh-THEE-uh). Felicia; blue daisy.
Agathis (AG-uh-thiss). Dammar pine, kauri pine.
Agave (uh-GAY-vee; uh-GAH-vee). Century plant.
Ageratum (uh-JER-uh-tum; aj-ur-AY-tum). Annual bedding plants.
Aglaonema (ag-lay-o-NEE-muh). Greenhouse foliage plants.
Agrarius (uh-GRAYR-ee-us). Of the fields.
Agrestis (uh-GRESS-tiss). Growing wild.
Agrostemma (ag-ro-STEM-uh). Corn cockles.
Agrostis (uh-GROS-tiss). Bent, cloud, redtop grasses.
Ailanthus (ay-LAN-thus). Tree of heaven.
Ajania (uh-JAN-ee-uh). Perennial herbs and shrubs of Ajan, East Africa.

Ajuga (AJ-yoo-guh; a-JYOO-guh). Ground cover plants; bugle weed.
Akebia (uh-KEE-bee-uh). Asian woody vines.
Alata (uh-LAY-tuh). Winged.
Alba (AL-buh). White.
Albescens (al-BES-enz). Whitish, becoming white.
Albicans (AL-bik-anz). Whitish.
Albicaulis (al-bih-KAW-liss). White-stemmed.
Albida (AL-bid-uh). White.
Albiflorus (al-bih-FLO-rus). White-flowered.
Albifrons (AL-bih-fronz). White-fronded.
Albizia (al-BIZ-ee-uh). Acacia-like trees, shrubs; silk tree.
Albomarginata (AL-bo-mar-jin-AY-tuh). White-margined.
Albuca (al-BYOO-kuh). White-colored (flowers).
Alcea (al-SEE-uh; AL-see-uh). Hollyhock.
Alchemilla (al-keh-MIL-uh). Lady's mantle.
Alcicornis (al-sih-KOR-niss). Elk-horned.
Aletris (AL-eh-triss). Star grass, blazing star.
Aleurites (al-yoo-RY-teez). Tung-oil tree, varnish tree.
Alexandrae (al-ex-AN-dree). From Alexandria, Egypt.
Alga (AL-guh). Elementary aquatic plants.
Allamanda (al-uh-MAN-duh). Tropical, woody flowering vines.
Allegheniensis (al-leh-GEH-nee-EN-siss). Growing in the Alleghenies.
Allium (AL-ee-um). Onion, garlic, chives.
Alluaudia (a-loo-OW-dee-uh; a-loo-AW-dee-uh). Succulent shrubs or trees.
Alnifolia (al-nih-FO-lee-uh). Alder-leaved.
Alnus (AL-nus). Trees and shrubs known as alder.
Alocasia (al-o-KAY-zhi-uh; al-o-KAY-zhuh; al-o-KAY-zi-uh). Greenhouse foliage plants.
Aloe (AL-o-ee; AL-o as common name). African succulents.
Alonsoa (uh-LON-zo-uh; a-lon-ZO-uh). Flowering greenhouse plants.
Alopecurus (a-lo-peh-KYUR-us). Fox-tail.

Aloysia (a-LOY-zee-uh; al-o-ISS-ee-uh). From proper name.
Alpestris (al-PESS-triss). Nearly alpine.
Alpinia (al-PIN-ee-uh). Shellflower.
Alpinus (al-PY-nus). Alpine.
Alsophila (al-SOF-ih-luh). Tree ferns.
Alstroemeria (al-streh-MEE-ree-uh). Peruvian lily.
Alternanthera (al-tur-NAN-thur-uh). Carpet bedding plants.
Alternifolia (al-ter-nih-FO-lee-uh). Alternate-leaved.
Althaea (al-THEE-uh). Hollyhock.
Altissima (al-TISS-ih-muh; awl-TISS-ih-muh). Tallest.
Altus (AL-tus). Tall.
Alyssum (uh-LISS-um). Rock garden plants.
Amabilis (uh-MAB-uh-liss). Lovely.
Amaranthus (am-uh-RAN-thus). Prince's-feather, Joseph's-coat.
Amarcrinum (am-ahr-CRY-num). Amaryllis-crinum hybrid.
Amarum (uh-MAR-um). Bitter.
Amaryllis (am-uh-RIL-iss). Belladonna lily.
Amauropelta (a-maw-ro-PEL-tah). Epiphytic or terrestrial ferns.
Amazonicus (a-mih-ZON-ih-kus). Of the Amazon river.
Ambiguus (am-BIG-yoo-us). Uncertain.
Amelanchier (am-eh-LANG-kee-ur). Juneberry, serviceberry, shadblow.
Americana (uh-mer-ih-KAY-nuh). From America.
Amethystinus (a-meh-THISS-tih-nus). Violet-colored.
Ammobium (a-MO-bee-um). Winged everlasting.
Ammophila (am-MOF-ih-luh). Sand-binding grasses.
Amoenus (uh-MEE-nus). Charming, pleasing.
Amomum (a-MO-mum). Referring to the acrid seeds of a plant; name of an East Indian genus, species of *Cornus*.
Amorpha (uh-MOR-fuh). False indigo.
Amorphophallus (uh-mor-fo-FAL-us). Immense tuberous herbs.
Ampelopsis (am-peh-LOP-siss). Woody vines.
Amphitecna (am-fih-TEK-nuh). With biocular ovaries.
Amplexicaulis (am-plek-sih-KAWL-iss). Stem clasping.
Amsonia (am-SO-nee-uh). Perennial border herbs.

Amurensis (a-moor-EN-siss). Of the Amur river region (East Siberia).
Amygdalus (uh-MIG-duh-lus). Peach, almond, flowering almond.
Anacardium (an-uh-KAHR-dee-um). Cashew.
Anagallis (an-uh-GAL-us). Pimpernel.
Ananas (uh-NAN-us; uh-NAY-nus; uh-NAH-nus). Pineapple.
Anaphalis (uh-NAF-uh-liss). Pearly everlasting.
Anceps (AN-seps). Two-headed, two-edged.
Anchusa (an-KYOO-sah). Alkanet, bugloss.
Andersoni (an-dur-SO-nee). From proper name.
Andinus (an-DEE-nus). Of the Andes.
Andraeanum (an-dree-AY-num). From proper name.
Andromeda (an-DROM-eh-duh). Bog rosemary, common name for *Pieris*.
Andropogon (an-dro-PO-gon). Beard grass.
Androsace (an-DROSS-uh-see). Rock jasmine.
Androstephium (an-dro-STEE-fee-um). Prairie bulbous plants.
Anemone (uh-NEM-o-nee). Windflower.
Anemonella (uh-nem-o-NEL-uh). Rue anemone.
Anethum (uh-NEE-thum). Dill.
Angelica (an-JEL-ih-kuh). Bold herbs of carrot family.
Angelonia (an-jeh-LO-nee-uh). Greenhouse flowering plants.
Angiospermous (an-jee-o-SPUR-mus). Seeds within pericarp.
Anglicus (ANG-glih-kus). English, of England.
Angraecum (an-GREE-kum). Tree-perching orchids.
Angulatus (an-gyoo-LAY-tus). Angular, angled.
Angustifolius (an-gus-tih-FO-lee-us). Narrow-leaved.
Anigozanthos (an-ih-go-ZAN-thus). Kangaroo's paw, Australian sword lily.
Anisodontea (an-iss-o-DON-tee-uh; a-ny-zo-DON-tee-uh). Irregularly-toothed (leaves).
Annona (a-NO-nuh). Tropical fruit trees; cherimoya.
Annularis (an-yoo-LAY-rus). Ring-shaped.
Annuus (AN-yoo-us). Annual.

Anomalus (a-NOM-uh-lus). Unusual.
Anopetalus (an-o-PET-uh-lus). Erect-petaled.
Antarctica (ant-AHRK-tik-uh). Of the Antarctic.
Antennaria (an-teh-NAY-ree-uh). Everlasting, pussytoes.
Anthemis (AN-theh-miss). Camomile.
Anthericum (an-THER-ih-kum). St.-Bernard's-lily.
Anthracnose (an-THRAK-nohs). Plant diseases.
Anthriscus (an-THRISS-kus). Chervil.
Anthurium (an-THYOO-ree-um). Greenhouse plants.
Antigonon (an-TIG-o-non). Coral vine, *Rosa montana*.
Antirrhinum (an-tih-RY-num). Snapdragon.
Aperta (uh-pur-tuh). Open.
Apetalous (ay-PET-al-us). Lacking petals.
Aphananthe (a-fuh-NANTH-ee). With inconspicuous flowers.
Aphelandra (af-eh-LAN-druh). Showy greenhouse plants.
Aphid (AY-fid; AF-id). Plant louse; common name.
Aphis (AY-fiss; AF-iss). Genus of plant lice.
Aphylla (uh-FIL-uh). Without leaves.
Apiculata (ay-pik-yoo-LA-tuh). Short, pointed tip.
Apiifolia (a-pee-ih-FO-lee-uh). Celery-leaved.
Apios (AY-pee-os; AP-ee-os). Wild or potato bean, groundnut.
Apium (AY-pee-um; AP-ee-um). Celery.
Apocynum (uh-POS-ih-num). Dogbane, Indian hemp.
Aprica (AP-ree-kuh). Exposed to the sun.
Aptenia (ap-TEEN-ee-uh; ap-TEEN-yuh). Wingless (seed capsule).
Aquatica (uh-KWAT-ih-kuh). In or near water.
Aquifolium (a-kwih-FO-lee-um). Holly-leaved.
Aquilegia (ak-wih-LEE-jee-uh). Columbine.
Arabica (uh-RAB-ih-kuh). From Arabia.
Arabis (AYR-uh-biss). Rock cress.
Arachis (AYR-uh-kuss). Spider-like.
Aralia (uh-RAY-lee-uh). Devil's-walking-stick, Hercules'-club.
Araucaria (ar-aw-KAY-ree-uh). Norfolk Island pine, monkey puzzle, etc.
Arborea (ahr-BO-ree-uh). Tree-like, woody.

Arborescens (ahr-bo-RES-enz). Tree-like, woody.
Arboretum (ahr-bo-REE-tum). Garden of trees and shrubs.
Arborvitae (AHR-bor-VY-tee). *Thuja*; coniferous evergreens.
Arbutifolia (ar-byoo-tih-FO-lee-uh). Arbutus-leaved.
Arbutus (ahr-BYOO-tus). Broad-leaved evergreen trees.
Archontophoenix (ahr-kon-toh-FEE-niks). King palm.
Arctostaphylos (ark-toh-STAF-ih-los). Bearberry.
Arctotis (ark-TOH-tiss). African daisy.
Arcuatus (ahr-kyoo-AY-tus). Arched.
Ardisia (ahr-DIZ-ee-uh). Red-berried Christmas plant.
Areca (a-REE-kuh; AR-ee-kuh). Decorative palms.
Arecastrum (a-rih-KASS-trum). Queen's palm.
Arenaria (a-reh-NAY-ree-uh). Sandwort; herb of pink family.
Arenarius (a-reh-NAY-ree-us). Of sandy places.
Arenga (ahr-EN-guh). From Malayan name for this palm.
Areolatus (uh-ree-o-LAYT-us). Pitted or netted.
Argemone (ahr-JEM-o-nee). Prickly poppy.
Argentea (ahr-JEN-tee-uh). Silvery.
Argutus (ahr-GYOO-tus). Sharp-toothed.
Argyraea (ahr-jih-REE-uh). Silvery.
Argyranthemum (ahr-jer-ANTH-ih-mum). Silver-flowered.
Argyrites (ahr-jih-RY-teez). With silver specks.
Arisaema (ar-ih-SEE-muh). Jack-in-the-pulpit.
Aristatus (ar-riss-TAYT-us). Bearded.
Aristea (a-RISS-tee-uh). With bristle-like, pointed (leaves).
Aristolochia (a-riss-toh-LO-kee-uh). Dutchman's-pipe, gooseflower.
Aristotelia (a-riss-toh-TEE-lee-uh). From proper name.
Armeria (ahr-MEH-ree-uh). Thrift; now statice.
Armoracia (ahr-mor-AY-see-uh). Horseradish.
Arnica (AHR-nee-kuh). Perennial medicinal herbs.
Aronia (uh-RO-nee-uh). Chokeberry; deciduous shrubs.
Aromatica (a-ro-MAT-ih-kuh). Aromatic.
Arrhenatherum (ayr-eh-nuh-THER-rum). With bristled, staminate flowers.

Artemisia (ahr-teh-MIZ-ee-uh; ahr-teh-MEE-zhee-uh). Wormwood, sagebrush.
Articulatus (ahr-tik-yoo-LAYT-us). Articulated, jointed.
Artocarpus (ahr-toh-KAHR-pus). Breadfruit.
Arum (AY-rum; AYR-um as common name). Black calla.
Aruncus (uh-RUN-kus). Spirea-like herbs; goatsbeard.
Arundinaria (uh-run-dih-NAY-ree-uh). Bamboo-like grasses; Southern cane.
Arundo (uh-RUN-doh). Tall grasses; true reeds.
Arvensis (ahr-VEN-siss). Of cultivated fields.
Asarum (ASS-ah-rum). Wild ginger.
Asclepias (ass-KLEE-pee-us). Milkweed, butterfly weed.
Ascyrum (a-SY-rum). St.-Andrew's-cross.
Asiatica (ay-zhee-AT-ih-kuh). Of Asia.
Asimina (uh-SIM-ih-nuh). Pawpaw.
Aspericaulis (ass-pur-ih-KAW-liss). Rough-stemmed.
Asperifolia (ass-pur-ih-FO-lee-uh). Rough-leaved.
Asperifolius (ass-pur-ih-FO-lee-us). Rough-leaved.
Asperula (ass-PUR-yoo-luh). Woodruff.
Asphodeline (ass-fod-eh-LY-nee). Jacob's-rod; also asphodel.
Aspidistra (ass-pih-DISS-truh). Foliage plant; cast-iron plant.
Asplenifolia (ass-plee-nih-FO-lee-uh). Leaf like a fern frond.
Asplenium (ass-PLEE-nee-um). Spleenwort fern.
Assurgens (ass-SUR-jenz). Ascending.
Aster (ASS-tur). Perennial herbs; Michaelmas daisy.
Asteriscus (ass-ter-ISS-cuss). Little star.
Astilbe (uh-STIL-bee). Perennials often called false spirea.
Astragalus (ass-TRAG-uh-lus). Milk vetch, locoweed.
Astrantia (ass-TRAN-chee-uh). Masterwort.
Astrocaryum (ass-tro-KAY-ree-um). Mexican feather palm.
Astrophytum (ass-tro-FY-tum). Mexican cactus.
Atamasco (at-uh-MASS-ko). Any zephyranthes.
Athrotaxis (ath-ro-TAK-siss). With a crowded arrangement (leaves).
Athyrium (a-THEER-ee-um). Ferns.

Atriplex (AT-rih-pleks). Saltbush, greasewood.
Atrorubens (a-tro-ROO-benz). Dark red.
Atropa (AT-ro-puh). Poisonous herbs; belladonna.
Atropurpurea (at-ro-pur-PYUR-ree-uh). Dark purple.
Atrosanguineus (at-ro-san-GWEEN-ee-us). Dark blood-red.
Atrovirens (at-ro-VY-renz). Dark green.
Attenuata (at-ten-yoo-AY-tuh). Slenderly tapering.
Aubrietia (aw-BREE-she-uh; aw-BREE-shuh). Perennials forming mats.
Aucuba (aw-KYOO-buh). Greenhouse foliage plants; Asiatic evergreen shrubs.
Augustifolium (aw-gus-tih-FO-lee-um). Majestic foliage.
Aurantiacus (aw-ran-tee-AY-kus). Orange-red.
Aureus (AW-ree-us). Golden.
Auricula (aw-RIK-yoo-luh). Early name for some primulas.
Auriculata (aw-rik-yoo-LAY-tuh). Ear-shaped appendage.
Australis (aws-TRAY-liss; aws-TRAW-liss). Southern.
Autumnale (aw-tum-NAY-lee). Autumnal.
Avena (uh-VEE-nuh). Oats.
Averrhoa (a-ver-O-uh). Ornamental trees; from proper name.
Avicularis (uh-vik-yoo-LAY-riss). Pertaining to birds.
Axil (AK-sil). Juncture or angle of leaf and stem.
Axillary (AK-sih-lehr-ee). In an axil.
Azalea (uh-ZAY-lee-uh; uh-ZAYL-yuh as common name). Showy shrubs, botanically *Rhododendron*.
Azara (uh-ZAW-ruh). Chilean evergreen plants.
Azorella (ay-zor-EL-luh). Mosquito fern, water fern.
Azureus (a-ZYOO-ree-us). Azure, sky-blue.

B

Babiana (bab-ee-AY-nuh). Low bulbous herbs.
Babylonica (bab-ih-LON-ih-kuh). From Babylon.
Baccata (ba-KAY-tuh). Berried, or berry-like.
Baccharis (BAK-uh-riss). Groundsel bush.

Bacopa (bay-KO-puh, buh-KO-puh). Water hyssop.
Baeria (BAY-ree-uh). California annuals; goldfields.
Baileya (BAY-lee-uh). From proper name.
Ballota (ba-LO-tuh). Black horehound.
Balsam (BAWL-sam). Impatiens; garden annuals.
Balsamifera (bal-sam-IF-er-uh). Bearing balsam.
Bamboo (bam-BOO). Giant woody grasses.
Bambusa (bam-BYOO-suh). Genus of typical bamboos.
Banksia (BANGK-see-uh). Australian evergreen trees, shrubs.
Baptisia (bap-TIZ-ee-uh; bap-TIZH-ee-uh). Wild or false indigo.
Barbarea (bahr-buh-REE-uh). Weedy herbs; winter cress.
Barbatus (bahr-BAY-tus). Barbed, bearded.
Barbellate (BAHR-beh-layt). Finely barbed.
Barclayana (bahr-klay-A-nuh). From proper name.
Barleria (bahr-LEE-ree-uh). Greenhouse shrubby herbs.
Bauera (BOW-ur-uh; BOW-ee-ruh). Cool greenhouse plants.
Bauhinia (baw-HIN-ee-uh). From proper name; tropical trees, shrubs, vines.
Beaumontia (bo-MON-she-uh; bo-MON-tee-uh). Herald's-trumpet.
Begonia (bee-GO-nee-uh). Greenhouse, bedding and house plants.
Begoniaceae (bee-go-nee-AY-see-ee). Begonia family.
Belamcanda (bel-am-KAM-duh). Blackberry lily.
Belladonna (bel-uh-DON-uh). Beautiful woman; atropa.
Bellis (BEL-iss). English daisy.
Bellus (BEL-us). Handsome.
Belmoreana (bel-mo-ree-AY-nuh). From proper name.
Beloperone (bel-o-pur-O-nee). Shrimp plant.
Benzoin (BEN-zo-in; BEN-zoin). Spicebush, species name of *Lindera*.
Berberis (BUR-bur-iss). Barberry.
Berchemia (ber-KEE-mee-uh). Woody vines.
Bergenia (bur-GEN-ee-uh). Perennials like saxifrages.

Bertolonia (bur-toh-LO-nee-uh). Greenhouse foliage plants.
Beschorneria (beh-shor-NEH-ree-uh). From proper name.
Beta (BEE-tuh). Beet.
Betula (BET-yoo-luh). Birch.
Bicolor (BY-kuh-lur; rarely BIK-o-lur). Two-colored.
Bicornis (by-KOR-niss). Two-horned.
Bicuspidata (by-kus-pih-DAY-tuh). Tipped with two rigid points.
Bidens (BY-denz). Sticktight, tickseed.
Biennial (by-EN-ee-al). Living two years, flowering the second.
Bifida (BIFF-ih-duh). Twice cut or cleft.
Biflora (by-FLOR-uh). Double-flowered.
Bifurcatus (by-fur-KAY-tus). Twice forked.
Bigeneric (by-jeh-NEHR-ik). Of two genera.
Bignonia (big-NO-nee-uh). Cross vine; trumpet vine is *Campsis*.
Billbergia (bil-BUR-jee-uh). Tropical epiphytic herbs.
Biloba (by-LO-buh). Double-lobed.
Biota (by-O-tuh). Form of Oriental arborvitae.
Bipinnata (by-pin-AY-tuh). Twice pinnate.
Bipinnatifida (by-PIN-uh-TIF-ih-duh). Double-pinnate.
Bistorta (bih-STOR-tuh). Double-twisted.
Blandus (BLAN-dus). Bland, mild.
Blechnum (BLEK-num). Rather coarse ferns; saw, deer, etc.
Boltonia (bohl-TOH-nee-uh). Tall aster-like perennials.
Bombax (BOM-baks). Silk-cotton tree.
Borago (bor-AY-go). Borage.
Borbonica (bor-BON-ee-kuh). Of Bourbonne, France.
Borealis (bor-ee-AY-liss). Northern.
Boronia (bor-O-nee-uh). Flowering greenhouse plants.
Bothriochloa (bo-three-OK-low-uh). Pitted grass.
Botrychium (bo-TRIK-ee-um). Hardy wild ferns.
Botryoides (bot-ree-OY-deez; boh-tree-OY-deez). Like a cluster of grapes; species of *Muscari*.
Botrytis (bo-TRY-tiss). Fungal disease.
Bougainvillea (boo-gin-VIL-lee-uh). Showy tropical vines.
Bouteloua (boot-uh-LOO-uh). Grama grass.

Bouvardia (boo-VAHR-dee-uh). Showy greenhouse shrubs.
Bowringiana (bo-ring-ee-AY-nuh). From proper name.
Boykinia (boy-KIN-ee-uh). California perennial herbs.
Brachyanthum (brak-ee-AN-thum). Short-flowered.
Brachycalyx (brak-ee-KAY-liks). Having a short calyx.
Brachycera (brah-KISS-eh-ruh). Short-horned.
Brachychiton (brak-ee-KY-ton). With short seed coverings.
Brachycome (brah-KIK-o-mee). Swan River daisy.
Brachyglottis (brak-ee-GLOT-iss). Short-tongued (florets).
Brachyloba (brak-ee-LO-buh). Short-lobed.
Brachypetalus (brak-ih-PET-uh-lus). Short-petaled.
Brachyphylla (brak-ee-FIL-luh). Short-leaved.
Bract (BRAKT). Modified leaf below flower.
Bracteatus (brak-tee-AY-tus). Bearing bracts.
Brahea (BRAY-hee-uh). Mexican fan palm.
Brasenia (bruh-SEE-nee-uh). Water shield.
Brassavola (brass-uh-VO-luh). Tropical American orchids.
Brassia (BRASS-ee-uh). Spider orchid.
Brassica (BRASS-ih-kuh). Cabbage, cauliflower, mustard, turnip.
Brassocattalelia (brass-o-kat-uh-LEE-lee-uh). Trigeneric orchid hybrids.
Brassocattleya (brass-o-KAT-lee-uh). Hybrids of *Brassavola* and *Cattleya*.
Brassolaelia (brass-o-LEE-lee-uh). Hybrids of *Brassavola* and *Laelia*.
Braziliensis (bruh-zil-ee-EN-siss). Of Brazil.
Brevibracteata (BREV-ih-brak-tee-AY-tuh). Short-bracted.
Brevicaulis (brev-ih-KAWL-iss). Short-stemmed.
Brevicornu (brev-ih-KOR-nyoo). Short-horned.
Breviflora (brev-ih-FLOR-uh). Short-flowered.
Brevifolius (brev-ih-FO-lee-us). Short-leaved.
Breviloba (brev-ih-LO-buh). Short-lobed.
Brevipes (BREV-ih-peez). Short-footed or short-stalked.
Breviscapus (brev-iss-KAY-pus). Short-stemmed.
Brevoortia (breh-VOR-tee-uh). California bulbous plant.

Brilliantissimus (bril-yan-TISS-ih-mus). Very brilliant.
Brittanicus (brih-TAN-ih-kus). Of Britain.
Briza (BRIH-zuh). Quaking grass.
Brodiaea (bro-dih-EE-uh). Showy California bulbous herbs.
Bromelia (bro-MEE-lee-uh). Tropical pineapple-like herbs.
Bromus (BRO-mus). Brome grass, also rescue, quake, etc.
Broussonetia (broo-so-NESH-ee-uh). Paper mulberry.
Browallia (bro-WAL-ee-uh). Flowering bedding plants.
Bruckenthalia (bruk-en-THAY-lee-uh). Spike heath.
Brunfelsia (brun-FEL-see-uh; brun-FEL-zee-uh). Lady-of-the-night.
Brunnera (brun-NEE-ruh). From proper name.
Bryonia (bry-O-nee-uh). Robust vines; bryony.
Bryophyllum (bry-o-FIL-um). Air plant, life plant.
Buddleia (bud-LEE-uh; BUD-lee-uh). Butterfly bush.
Bulbiferous (bulb-IF-ur-us; bul-BIF-ur-us). Bearing bulbs.
Bulbine (bul-BY-nee). Bulbous.
Bulbocodium (bul-bo-KO-dee-um). Crocus-like herbs.
Bulbophyllum (bul-bo-FIL-um). Old World orchids.
Bulbous (BUL-bus). Having bulbs.
Bullata (bul-LAY-tuh). Puckered, blistered.
Bumelia (byoo-MEE-lee-uh). Thorned shrubs; false buckthorn.
Buphthalmum (byoof-THAL-mum). Oxeye.
Bursera (bur-SUR-uh). From proper name.
Butia (BYOO-tee-uh). Fairly hardy palms.
Buxifolius (buk-sih-FO-lee-us). Box-leaved.
Buxus (BUK-sus). Boxwood, box.
Byroides (bry-OY-deez). Moss-like.

C

Cabomba (kuh-BOM-buh). Commonest aquarium plant.
Cactaceae (kak-TAY-see-ee). Cactus family.
Cacti (KAK-ty). Plural of cactus.
Cactus (KAK-tus). Spiny desert plants.

Caduceous (kuh-DYOO-see-us). Falling early.
Caeruleus (seh-ROO-lee-us; see-ROO-lee-us). Cerulean, dark blue.
Caesalpinia (sez-al-PIN-ee-uh; ses-al-PIN-ee-uh). Brazilwood.
Caesius (SEE-zee-us). Bluish-gray.
Cajanus (kuh-JAY-nus). Pigeon pea, cajan.
Calacinum (kal-uh-SY-num). Wire plant or vine.
Caladium (kuh-LAY-dee-um). Greenhouse foliage plants.
Calamagrostis (kal-am-uh-GRAW-stiss). Reed grass.
Calamintha (ka-luh-MIN-thuh). Beautiful mint.
Calamus (KAL-uh-mus). Rattan and cane palm.
Calandrinia (kal-an-DRIN-ee-uh). Rock purslane.
Calanthe (kuh-LAN-thee). Showy tropical orchids.
Calathea (kal-uh-THEE-uh). Warmhouse foliage plants.
Calathinus (kal-uh-THY-nus). Basket-like.
Calceolaria (kal-see-o-LAY-ree-uh; kal-see-o-LAYR-ee-uh as common name). Florists' plants.
Calendula (kuh-LEN-dyoo-luh). Pot marigolds.
Calendulacea (ka-len-dyoo-LAY-see-uh). Species of azalea.
Calibrachoa (kal-ib-brah-KO-uh). From proper name.
Calla (KAL-uh). Water arum, wild calla.
Calliandra (kal-ee-AN-druh). Showy tropical trees and shrubs.
Callicarpa (kal-ih-KAHR-puh). Beautyberry, French mulberry.
Callirrhoe (ka-LIHR-o-ee). Poppy mallow.
Callistemon (kal-ih-STEE-mon). Bottle brush.
Callistephus (ka-LISS-tih-fus). China or garden aster.
Callosa (kal-O-suh). Thick- or hard-skinned.
Calluna (ka-LOO-nuh). Heather.
Calocephalus (kal-o-SEFF-uh-lus). Foliage edging plants.
Calochortus (kal-o-KOR-tus). Mariposa lily.
Calodendrum (kal-o-DEN-drum). Beautiful tree.
Calonyction (kal-o-NIK-tee-on). Moonflower.
Calophyllum (kal-o-FILL-um). Aromatic tropical trees.
Calopogon (kal-o-PO-gon). Bog orchids; grass pink.
Caltha (KAL-thuh). Marsh marigold.

Calycanthus (kal-ih-KAN-thus). Sweet shrub, strawberry shrub.
Calycinus (kal-ih-SY-nus). Calyx-like.
Calyculata (kuh-LIK-yoo-LAH-tuh). Bracts around calyx.
Calylophus (ka-lih-LO-fuss). From proper name.
Calypso (kuh-LIP-so). Hardy bog orchid.
Calyx (KAY-liks). Outermost case of flower.
Camassia (kuh-MAS-ee-uh). American bulbous herbs.
Cambium (KAM-bee-um). Growing tissue under bark.
Camellia (kuh-MEEL-ee-uh; kuh-MEEL-yuh). Evergreen flowering shrubs.
Camissonia (kam-ih-SO-nee-uh). From proper name.
Campanula (kam-PAN-yoo-luh). Bellflower, Canterbury bells.
Campanulaceae (kam-pan-yoo-LAY-see-ee). Bellflower family.
Campanulate (kam-PAN-yoo-layt). Bell-shaped, cup-shaped.
Campestris (kam-PESS-triss). Of the fields or plains.
Campsidium (kamp-SID-ee-um). Southern evergreen climbers.
Campsis (KAMP-siss). Trumpet vine or creeper.
Camptosorus (kamp-toh-SO-rus). Walking fern.
Canadensis (kan-a-DEN-siss). From Canada.
Canaliculatus (kan-uh-LIK-yoo-lay-tus). Channeled, grooved.
Candicans (KAN-dih-kanz). White, hoary.
Candidissimus (kan-dih-DISS-ih-mus). White hairy or hoary.
Candidum (KAN-dih-dum). Pure white, shining.
Canella (ka-NEL-uh). Wild cinnamon.
Canescent (kuh-NESS-ent). With gray pubescence.
Canina (kuh-NY-nuh). Pertaining to a dog.
Canistrum (kuh-NISS-trum). Showy conservatory plants.
Canna (KAN-uh). Tropical bedding plants.
Cannabis (KAN-uh-biss). Hemp.
Canus (KAY-nus). Gray- or ash-colored.
Capensis (kuh-PEN-siss). Of the Cape of Good Hope.
Capillaris (ka-pihl-LAY-riss). Hair-like.
Capillus Veneris (kuh-PIL-us VEN-ur-us). Hair of Venus.
Capitata (kap-ih-TA-tuh). Like a head.
Caprea (KAP-ree-uh). Pertaining to a goat.

Capsicum (KAP-sih-kum). Peppers; garden, green, red, etc.
Capsule (KAP-syool). Dry, splitting fruit.
Caragana (kayr-uh-GAY-nuh). Pea shrub or tree.
Cardamine (kahr-DAM-ih-nee). A kind of cress.
Cardinalis (kahr-dih-NAY-liss). Cardinal.
Cardiospermum (kahr-dee-o-SPUR-mum). Balloon vine.
Carex (KAY-reks). Sedge; grass-like plants.
Carica (KA-rih-kuh). Papaya; tropical fruit.
Carinata (kar-ih-NAT-uh). Having a keel.
Carissa (kuh-RISS-uh). Tropical hedge plants; karanda.
Carludovica (kahr-loo-doh-VY-kuh). Panama hat plant.
Carminatus (kahr-mih-NAY-tus). Carmine.
Carnation (kahr-NAY-shun). *Dianthus*.
Carnegiea (kahr-NAY-gee-uh). Giant cactus or saquaro.
Carneus (KAHR-nee-us). Flesh-colored.
Carnosus (kahr-NO-sus). Fleshy.
Carolinensis (ka-ro-ly-NEN-siss). Growing in the Carolinas.
Carolinianus (ka-ro-lin-ee-AY-nus). From the Carolinas.
Carota (kuh-RO-tuh). Latin for carrot.
Carpel (KAHR-pel). Simple pistil.
Carpenteria (kahr-pen-TEE-ree-uh). *Philadelphus*-like shrub.
Carpinus (kahr-PY-nus). Hornbeam.
Carpobrotus (kahr-po-BRO-tuss). Having edible fruit (some species).
Cartilagineus (kahr-tih-LAJ-ih-nus). Like cartilage.
Carum (KAY-rum). Caraway.
Carya (KAY-ree-uh; KAHR-ee-uh). Hickory.
Caryophyllus (kayr-ee-o-FIL-us). Pertaining to clove.
Caryopteris (kar-ee-OP-tur-iss). Blue spirea, bluebeard.
Caryota (kar-ee-O-tuh). Fishtail palm.
Casimiroa (kass-ih-mih-RO-uh). White sapote, Mexican apple.
Caspius (KASS-pee-us). Caspian.
Cassandra (ka-SAN-druh). Synonym of *Chamaedaphne*.
Cassia (KASH-ee-uh; KASS-ee-uh). Senna, coffee senna.
Cassinoides (kass-ih-NOY-deez). Cassine-like.

Cassiope (ka-SY-o-pee). Evergreen ericaceous shrubs.
Castanea (kass-TAY-nee-uh). Chestnut.
Castanopsis (kass-tuh-NOP-siss). Chinquapin.
Castilla (kass-TIL-uh; kass-TIL-yuh). Mexican rubber tree.
Castilleja (kass-tih-LEE-yuh). Painted cups.
Casuarina (ka-szhur-EYE-na). Resembling a cassowary.
Catalpa (kuh-TAL-puh). Hardy flowering trees.
Catananche (kat-uh-NAN-kee). Cupid's-dart, blue succory.
Catasetum (kat-uh-SET-um). Tropical American orchids.
Catesbaea (kats-BEE-uh). Lily thorns.
Cathartica (ka-THAR-tih-kuh). Purging.
Catkin (KAT-kin). Scaly spike, usually pendulous.
Cattleya (KAT-lee-uh). Commonest florists' orchids.
Caucasicus (kaw-KASS-ih-kus). Belonging to the Caucasus.
Caudata (kaw-DAY-tuh). Tailed.
Caudex (KAW-deks). Persistent base of herbaceous stem.
Caulescent (kaw-LESS-ent). Having a stem.
Caulophyllum (kaw-lo-FIL-um). Blue cohosh.
Ceanothus (see-uh-NO-thus). New Jersey tea, buckbrush.
Cecropia (see-KRO-pee-uh). Snakewood tree.
Cedrela (seh-DREE-luh; SED-ree-luh). Spanish cedar.
Cedrus (SEE-drus). True cedar.
Celastrus (see-LASS-trus). Bittersweet.
Celosia (see-LO-she-uh; see-LO-see-uh). Cockscomb.
Celtis (SEL-tiss). Hackberry.
Centaurea (sen-taw-REE-uh; sen-TAW-ree-uh). Cornflower, bachelor's-buttons.
Centranthus (sen-TRAN-thus). Red valerian, Jupiter's-beard.
Cephalanthus (sef-uh-LAN-thus). Buttonbush.
Cephalaria (sef-uh-LAY-ree-uh). Coarse annuals, perennials.
Cephalatus (sef-uh-LAY-tus). Bearing heads.
Cephalocereus (sef-uh-lo-SEE-ree-us). Columnar cacti; old-man cactus.
Cephalophyllum (sef-uh-lo-FIL-um). With head-shaped leaves.
Cephalotaxus (sef-uh-lo-TAK-sus). Yew-like evergreens.

Cephalotes (sef-uh-LO-teez). Like a small head.
Cerasifera (ser-uh-SIF-ur-uh). *Cerasus* or cherry-bearing.
Cerastium (seh-RAS-tee-um). Snow-in-summer, chickweed.
Cerasus (SER-uh-sus). A name for cherry.
Ceratonia (ser-uh-TOH-nee-uh). Carob, trees of Mediterranean region.
Ceratophyllum (ser-uh-toh-FIL-um). Aquarium plant; cedar moss.
Ceratopteris (ser-uh-TOP-tur-iss). Water or floating fern.
Ceratostigma (ser-at-o-STIG-muh). Plumbago, leadwort.
Cercidiphyllum (sur-sih-dih-FIL-um). Katsura tree.
Cercidium (sur-SID-ee-um). Paloverde.
Cercis (SUR-siss). Redbud, Judas tree.
Cercocarpus (sur-ko-KAHR-pus). Mountain mahogany.
Cereus (SEE-ree-us). Mostly columnar cactus.
Cerifera (seh-RIF-ur-uh). Wax-bearing.
Ceroxylon (see-ROKS-ih-lon). Wax palm.
Cespitosa (sess-pih-TOH-suh). In tufts; forming mats.
Cestrum (SESS-trum). Night and day jasmines.
Chaenomeles (kee-NOM-eh-leez). Japanese flowering quince.
Chalcedonica (chal-see-DON-ih-kuh). Of Chalcedon, Turkey.
Chamaecyparis (kam-eh-SIP-uh-riss). Decorative and timber conifers.
Chamaedaphne (kam-eh-DAF-nee). Bog shrub; leatherleaf.
Chamaedorea (kam-ee-DOR-ee-uh). With easily accessible fruit.
Chamaemelum (kam-uh-MEL-um). Chamomile.
Chamaerops (kuh-MEE-rops). Low fan palms.
Cheilanthes (ky-LAN-theez). Rock-loving ferns; lip ferns.
Cheiranthus (ky-RAN-thus). Wallflower.
Cheiridopsis (ky-rih-DOP-sis). Succulent perennials.
Chelidonium (kel-uh-DOH-nee-um). Greek word for swallow.
Chelone (kee-LO-nee). Turtlehead, snakehead.
Chenopodium (kee-no-PO-dee-um). Goosefoot, pigweed.
Chilopsis (ky-LOP-sis). Lip-like (calyx).

Chimaphila (ky-MAF-ih-la). Pipsissewa, prince's pine.

Chimonanthus (ky-mo-NAN-thus). Half-hardy deciduous shrubs from China.

Chimonobambusa (chih-mon-o-bam-BOO-suh). Winter bamboo.

Chincherinchee (chin-che-RIN-chee). *Ornithogalum thyrsoides.*

Chinensis (chin-EN-siss). Belonging to China.

Chiococca (kee-o-KO-kuh). With snow-white fruit.

Chionanthus (ky-o-NAN-thus). Fringe tree.

Chionodoxa (ky-on-o-DOK-suh). Glory-of-the-snow.

Chlorophyll (KLO-ro-fil). Green matter of plants.

Chlorophytum (klo-ro-FY-tum). Greenhouse foliage plants.

Chlorosis (klo-RO-siss). Loss of green color.

Choisya (SHWA-see-uh). Flowering evergreen shrub; from proper name.

Chorisia (kor-IZ-ee-uh). Floss silk tree.

Chorizema (ko-RIZ-eh-muh; kor-ih-ZEE-muh). Flowering pot plants.

Chromatella (kro-muh-TEL-uh). With color.

Chrysalidocarpus (kriss-o-lih-doh-KAHR-pus). With golden fruit.

Chrysanthemum (kriss-AN-thee-mum). Garden and greenhouse plants.

Chrysanthus (kriss-AN-thus). Golden-flowered.

Chrysobalanus (kriss-o-ba-LAY-nus). Coco plum.

Chrysogonum (kriss-SOG-o-num). Golden star, perennial herb.

Chrysophyllum (kriss-o-FIL-um). Star apple.

Chrysopsis (kriss-OP-siss). Golden aster.

Chrysothamnus (kris-o-THAM-nus). Golden shrub.

Cibotium (sy-BO-tee-um). Large florists' ferns.

Cichorium (sih-KO-ree-um). Chicory, endive.

Ciliata (sil-ee-A-ta). Margined with hairs; fringed.

Cimicifuga (sim-ih-SIF-yoo-guh). Bugbane, snakeroot.

Cineraria (sin-eh-RAY-ree-uh; sin-eh-RAYR-ee-uh). Forms of *Senecio cruentus.*

Cinnamomum (sin-uh-MO-mum). Cinnamon, camphor tree.

Circinata (sur-sih-NA-tuh). Coiled.
Cirrose (SIR-ohss). Bearing a tendril.
Cirsium (SUR-see-um). Thistle.
Cissus (SISS-us). Grape-like vines.
Cistus (SISS-tus). Rockrose.
Citharexylum (sith-uh-reks-EYE-lum). Fiddlewood, zitherwood.
Citrinus (sih-TRY-nus). Citron-colored or like.
Citrus (SIT-rus). Lemon, grapefruit, orange, etc.
Cladanthus (kluh-DAN-thus). Anthemis-like annuals.
Cladrastis (kluh-DRAS-tiss). Yellowwood tree.
Clandonensis (klan-duh-NEN-siss). Of Clandon, Surrey, UK.
Clarkia (KLAHR-kee-uh). Showy garden annuals.
Clavata (klay-VA-ta). Club-shaped.
Claytonia (klay-TOH-nee-uh). Spring beauty.
Cleistocactus (kly-sto-KAK-tuss). Closed (-flowered) cactus.
Clematis (KLEM-uh-tiss). Woody flowering vines.
Cleome (klee-O-mee). Popular annuals; spiderflower.
Clerodendron (klee-ro-DEN-dron). Showy woody vines and shrubs.
Clethra (KLETH-ruh; KLEE-thruh). Sweet pepper bush.
Clianthus (kly-AN-thus). Vine-like plants; glory pea.
Clintonia (klin-TOH-nee-uh). Woodland flowering herbs.
Clitoria (kly-TOH-ree-uh). Tropical vines; butterfly pea.
Clivia (KLY-vee-uh). Evergreen amaryllis-like plants.
Clone (KLOHN). Plants derived vegetatively from one specimen.
Clytostoma (kly-toh-STO-muh). Beautiful flowers.
Cneorum (nee-O-rum). Specific name of *Daphne*; low evergreen shrubs.
Cobaea (ko-BEE-uh). Cup-and-saucer vine.
Coccineus (kok-SIN-ee-us). Scarlet.
Coccinia (kok-SIN-ee-uh). Ivy gourd.
Coccoloba (ko-ko-LO-buh; kok-o-LO-buh). Sea grape, chickory grape.
Coccolobis (kok-o-LO-biss). Sea grape, seaside plum.

Cocculus (KOK-yoo-lus). Woody vines; Carolina moonseed.
Cochlearia (kok-lee-AY-ree-uh). Scurvy grass.
Cocos (KO-kos). Coconut palm.
Codiaeum (ko-dih-EE-um). Croton of trade.
Coelestis (see-LES-tiss). Celestial, sky-blue.
Coelogyne (see-LOJ-ee-nee). Asiatic epiphytic orchids.
Coerulea (seh-ROO-lee-uh; see-ROO-lee-uh). Dark blue.
Coffea (KOF-ee-uh). Trees or shrubs yielding coffee.
Coggygria (kog-ee-GREE-uh). Greek name for smoke tree.
Coix (KO-iks). Decorative grasses; Job's-tears.
Cola (KO-luh). African trees; kola nut.
Colchica (KOL-chih-kuh). From Colchis.
Colchicum (KOL-chih-kum; KOL-kih-kum). Autumn crocus.
Coleus (KO-lee-us). Showy foliage plants.
Colletia (ko-LEE-she-uh). Anchorplant.
Collinsia (ko-LIN-see-uh). Blue lips, blue-eyed Mary.
Collinsonia (ko-lin-SO-nee-uh). Horse balm, stoneroot.
Colocasia (kol-o-KAY-see-uh; kol-o-KAY-zee-uh). Elephant's-ear, dasheen.
Columbine (KOL-um-byn). *Aquilegia*.
Colutea (ko-LOO-tee-uh). Bladder senna.
Colvillea (kol-VIL-ee-uh). Tender African tree.
Commelina (ko-meh-LY-nuh). Plants like wandering Jew.
Communis (kom-MYOO-niss). Common, general, gregarious.
Comosus (ko-MO-sus). With long hair.
Compactus (kom-PAK-tus). Compact, dense.
Complexus (kom-PLEKS-us). Circled, embraced.
Compositae (kom-POZ-ih-tee). Daisy family.
Compressus (kom-PRESS-us). Compressed, flattened.
Comptonia (komp-TOH-nee-uh). Sweet fern, sweet bush.
Concinnus (kon-SIN-us). Well-made.
Concolor (KON-kul-ur). Of uniform color.
Confertus (kon-FUR-tus). Crowded; pressed together.
Confervae (con-FUR-vee). Pond scum alga.
Confusus (kon-FYOO-sus). Uncertain.

Conifer (KO-nih-fur; KON-ih-fur). Cone-bearing plant.
Coniferous (ko-NIF-ur-us). Cone-bearing.
Connate (KON-ayt). United congenitally, firmly.
Conocarpus (ko-no-KAHR-pus). With cone-shaped fruit.
Conoides (ko-NOY-deez). Cone-like.
Consolida (kon-SO-lee-duh). Firm, stable.
Conspicuus (kon-SPIK-yoo-us). Conspicuous.
Controversa (kon-tro-VUR-suh). Controversial.
Convallaria (kon-va-LAY-ree-uh). Lily of the valley.
Convolute (KON-vo-lyoot). Rolled longitudinally.
Convolvulus (kon-VOL-vyoo-lus). Bindweed, morning-glory.
Cooperia (koo-PEE-ree-uh). Rain or prairie lily.
Copallinus (ko-pal-LY-nus). Resinous, gummy.
Copernicia (ko-pur-NEE-see-uh). From proper name.
Coprosma (kop-ROS-muh). Tender foliage, hedge plants.
Coptis (KOP-tiss). Goldthread.
Corallinus (kor-ah-LY-nus). Coral red.
Corchorus (KOR-ko-rus). Plants supplying jute.
Cordate (kor-DAYT). Heart-shaped.
Cordifolius (kor-dih-FO-lee-us). With heart-shaped leaves.
Cordula (kor-DYOO-luh). Greenhouse lady's-slipper.
Cordyline (kor-dih-LY-nee). Plants resembling dracaenas.
Coreanus (kor-ee-AY-nus). Of Korea.
Coreopsis (ko-ree-OP-siss). Tickseed.
Coriaceous (ko-ree-AY-shus; kor-ee-AY-shus). Leathery.
Coriandrum (ko-ree-AN-drum; kor-ee-AN-drum). Coriander.
Coriaria (kor-ee-AYR-ee-uh). Leather (some species are used in tanning).
Corm (korm). Bulb-like, but solid.
Corniculatus (kor-nik-yoo-LAY-tus). Horned.
Cornus (KOR-nus). Dogwood or cornel.
Corolla (ko-ROHL-uh). Inner perianth, petals.
Corollatus (kor-o-LAY-tus). Corolla-like.
Corona (ko-ROH-nuh). Crown-like appendage on flower.

Coronarius (kor-o-NAY-ree-us). Of or belonging to a crown or garland, specific name of *Philadelphus*.
Coroniform (ko-RO-nih-form). Like a crown.
Coronilla (kor-o-NIL-uh). Scorpion senna, axseed.
Correa (KOR-ee-uh). Australian shrubs.
Corrugata (kor-yoo-GAT-uh). Wrinkled or in folds.
Cortaderia (kor-tuh-DEE-ree-uh). Pampasgrass.
Corydalis (ko-RID-uh-liss). Spring-flowering herbs.
Corylopsis (kor-ih-LOP-siss). Winter hazel.
Corylus (KOR-ih-lus). Hazelnut, filbert.
Corymb (KOR-imb; KOR-im). Flat-topped or convex cluster flowering inward.
Corypha (KOR-ih-fuh). Tropical fan palms.
Coryphantha (kor-ih-FAN-thuh). Small cylindrical cactus.
Cosmos (KOZ-mos; KOZ-mus). Showy late-blooming annuals.
Costata (kos-TAY-tuh). Ribbed.
Costus (KOSS-tuss). Aromatic plant.
Cotinus (KOT-in-us; ko-TY-nus). Smoke tree.
Cotoneaster (ko-TOH-nee-ASS-tur). Deciduous and evergreen shrubs.
Cotyledon (kot-ih-LEE-don). Old World succulents.
Cotyledon (kot-ih-LEE-dun). First leaf, as in bean.
Couroupita (kor-o-PEE-tuh). From the French Guianese name for this plant.
Crambe (KRAM-bee). Sea kale.
Crassifolius (krass-ih-FO-lee-us). Thick-leaved.
Crassipes (KRAS-sih-peez). Thick-footed or stalked.
Crassula (KRAS-yoo-luh). Succulent herbs and shrubs.
Crassus (KRAS-us). Thick, fleshy.
Crataegus (kruh-TEE-gus). Hawthorn, red haw.
Crateriform (kruh-TER-ih-form). Crater-like.
Crenata (kree-NAY-tuh). Toothed, teeth rounded.
Crenulatus (kren-yoo-LAY-tus). Finely scalloped.
Crepis (KREE-piss). Weedy herbs; hawk's-beard.
Crescentia (kre-SEN-she-uh; kre-SEN-tee-uh). Calabash tree.

Crinita (kry-NY-tuh). With long hairs.
Crinum (KRY-num). Lily-like amaryllids.
Crispatus (kriss-PAY-tus). Crisped, curled.
Crispus (KRISS-pus). Curled.
Cristata (kriss-TAY-tuh). Comb-like, crested.
Crocatus (kro-KAY-tus). Saffron-yellow.
Crocosmia (kro-KOZ-mee-uh). Bulbous gladiolus-like herb.
Crocus (KRO-kus). Early-flowering bulbous herb.
Crossandra (kro-SAN-druh). Showy greenhouse shrubs.
Crossopetalum (kross-o-PET-uh-lum). Fringe-leaved.
Crotalaria (kro-tuh-LAY-ree-uh; krot-uh-LAY-ree-uh). Southern cover crops.
Croton (KRO-ton). Florists' foliage plants; now *Codiaeum*.
Cruciferae (kroo-SIF-ur-ee). Mustard or cabbage family.
Cruciferus (kryoo-SIF-ur-us). Cross-bearing.
Cruentus (kryoo-EN-tus). Bloody.
Crustaceous (krus-TAY-shus). Hard and brittle.
Crustatus (krus-TAY-tus). Encrusted.
Cryptanthus (krip-TAN-thus). Florists' foliage plants.
Cryptogam (KRIP-toh-gam). Spore-bearing plant.
Cryptogamous (krip-TOG-a-mus). Pertaining to cryptogams.
Cryptogramma (krip-toh-GRAM-muh). Small alpine ferns.
Cryptomeria (krip-toh-MEE-ree-uh). Japanese evergreen tree.
Cryptostegia (krip-toh-STEE-jee-uh). Showy tropical woody vines.
Cucullate (KYOO-ku-layt; kyoo-KUL-ayt). Hooded.
Cucumis (KYOO-kyoo-miss). Cucumber, muskmelon.
Cucurbita (kyoo-KUR-bih-tah). Squash, pumpkin, gourd.
Cucurbitaceae (kyoo-kur-bih-TAY-see-ee). Cucumber family.
Culm (KULM). Stem of sedges, grasses.
Cuneate (KYOO-nee-ayt). Wedge-shaped.
Cuneatum (kyoo-nee-AY-tum). Wedge-shaped.
Cunninghamia (kun-ing-HAM-ee-uh). Asiatic evergreen trees.
Cuphea (KYOO-fee-uh). Cigar-flower.
Cupressocyparis (koo-PRESS-o-sy-PAR-iss). Leyland cypress.

Cupressus (kyoo-PRESS-us). Cypress; evergreen trees.
Cupreus (KYOO-pree-us). Copper-like or colored.
Curcuma (kur-KOO-mah). Yellow-colored.
Curviflora (kurv-ih-FLOR-uh). With curved flowers.
Cuscuta (koos-KOO-tuh). Dodder.
Cuspidata (kus-pih-DAY-tuh). Short, rigid point.
Cyananthus (sy-an-AN-thus). Blue-flowered.
Cyaneus (sy-AY-nee-us). Blue.
Cyanocarpus (sy-an-o-KAHR-pus). Blue-fruited.
Cyanotis (sy-uh-NO-tiss). Greenhouse creeping herb.
Cyanus (sy-AN-us). Blue.
Cyathea (sy-ATH-ee-uh). Tall tree ferns.
Cycadaceae (sik-uh-DAY-see-ee; sy-kuh-DAY-see-ee). Cycas or sago palm family.
Cycas (SY-kuss). Palm-like.
Cyclamen (SIK-luh-men; SY-kluh-men as common name). Potted plant.
Cyclanthera (sy-klan-THEE-ruh). Tender vines.
Cyclophorus (sy-KLOFF-o-rus). Felt, tongue or Japanese fern.
Cyclops (SY-klops). Cyclopean; gigantic.
Cydista (sy-DISS-tuh). Showy woody vines.
Cydonia (sy-DOH-nee-uh). Quince.
Cylindricus (sih-LIN-drih-kus). Cylindrical.
Cymbalaria (sim-buh-LAY-ree-uh). Kenilworth ivy.
Cymbidium (sim-BID-ee-um). Florists' spray orchids.
Cymbopogon (sim-bo-PO-gon). Aromatic perennial grasses.
Cyme (SIGHM). Flat-topped flower cluster.
Cymosus (si-MO-sus). Having a cyme or cymes.
Cynara (SIN-uh-ruh). Artichoke.
Cynodon (sin-O-don; SY-no-don). Bermuda grass.
Cynoglossum (sin-o-GLOSS-um; sy-no-GLOSS-um). Chinese forget-me-not.
Cyperus (sy-PEE-rus; sih-PEE-rus). Umbrella plant.
Cypress (SY-press). Southern evergreen trees; *Cupressus*.

Cypripedium (sip-rih-PEE-dee-um). Lady's-slipper, moccasin flower.
Cyrilla (seer-IH-luh). From proper name.
Cyrtomium (sur-TOH-mee-um). Holly fern.
Cystopteris (siss-TOP-tur-iss). Bladder fern.
Cytisus (SIT-ih-sus). Broom.

D

Daboecia (dab-EE-she-ah). Irish heath.
Dactylis (DAK-tih-liss). Orchard grass.
Dactylifera (dak-tih-LIFF-ur-ah). Finger-like.
Dactyloides (dak-tih-LOY-deez). Finger-like.
Daffodil (DAF-o-dil). Trumpet narcissus.
Dahlia (DAHL-yuh; DAL-yuh; DAYL-yuh). Tuberous garden plants.
Dahuricus (dah-HOO-rih-kus; dah-HOOR-ih-kus). Of Dahuria (Siberia).
Dalea (DAY-lee-uh). Desert shrubs.
Dalibarda (dal-ih-BAHR-duh). Dewdrop.
Dalmaticus (dal-MA-tih-kus). Dalmatian.
Damascenus (dam-uh-SEE-nus). Of Damascus.
Daphne (DAF-nee). Evergreen and deciduous shrubs.
Darlingtonia (dahr-ling-TOH-nee-uh). California pitcher plant.
Dasycalyx (das-ih-KAL-iks). With a hairy calyx.
Dasycarpus (das-ih-KAHR-pus). Thick-fruited.
Dasylirion (das-ih-LIR-ee-on). Tree-like desert plants; sotol.
Dasyphyllum (das-uh-FIL-um). Hairy- or shaggy-leaved.
Datura (duh-TYOO-ruh). Angel's-trumpet, Jimson weed.
Daucus (DAW-kus). Carrot.
Davallia (duh-VAL-ee-uh). Hare's-foot fern.
Davidia (day-VID-ee-uh). Dove tree.
Davidiana (day-vid-ee-AY-nuh). From proper name.
Dealbatus (dee-al-BAY-tus). Whitened, powdery.
Debilis (DEB-ih-liss). Weak, frail.

Decaisnea (de-KAYZ-nee-uh). Asiatic shrubs.
Deciduous (dee-SID-yoo-us). Not evergreen, leaves drop every autumn.
Decipiens (dee-SIP-ee-enz). Deceptive.
Declinatus (dek-lin-AY-tus). Bending down.
Decodon (DEK-o-don). Swamp loosestrife, water willow.
Decumaria (dek-yoo-MAY-ree-uh). Southern woody vines.
Decumbens (dee-KUM-benz). Reclining at the base, tips upright.
Decurrens (de-KUR-enz). Running down the stem.
Decussata (dee-ku-SAY-tuh). Alternate leaf pairs at right angles.
Deflexed (dee-FLEKST). Bent abruptly downward.
Dehiscent (dee-HISS-ent). Splitting along valves.
Delavayi (del-uh-VAY-ee-eye). From proper name.
Delicatus (del-ih-KAY-tus). Delicate, tender.
Deliciosa (dee-liss-ee-O-suh). Delicious.
Delonix (dee-LO-niks). Royal poinciana, peacock flower.
Delosperma (del-o-SPUR-muh). With visible seeds.
Delphinium (del-FIN-ee-um). Larkspur.
Deltoidea (del-TOID-ee-uh). Triangular.
Dendranthema (den-DRAN-theh-muh). Tree flower.
Dendrobium (den-DRO-bee-um). Showy spray orchids.
Dendrocalamus (den-dro-KAL-uh-mus). Tree-like grasses, giant bamboos.
Dendrology (den-DROL-o-jee). Study of trees.
Dendromecon (den-dro-MEE-kon). Bush or tree poppy.
Dennstaedtia (den-STET-ee-uh). Cup fern; hay-scented fern.
Densiflorus (den-si-FLO-rus). Densely flowered.
Dentata (den-TAY-tuh). Coarse-toothed.
Denticulata (den-tik-yoo-LAY-tuh). Toothed slighty or minutely.
Deodar (dee-o-DAHR). East Indian cedar.
Derris (DER-iss). Malay jewel vine.
Deschampsia (des-KAMP-see-uh; day-SHOMP-see-uh). Hair grass; from proper name.

Desmanthus (dez-MAN-thus). With bundled flowers.
Desmodium (des-MO-dee-um). Weedy herbs of pea family.
Deutzia (DYOOT-see-uh). Flowering hardy shrubs.
Diacanthus (dy-uh-KAN-thus). With pairs of spines.
Diandrus (dy-AN-drus). Having two stamens.
Dianella (dy-uh-NEL-luh). Flax lily.
Dianthus (dy-AN-thus). Pinks, carnation.
Diaphanus (dy-AF-uh-nus). Diaphanous, transparent.
Diascia (dy-ASH-ee-uh; dy-AS-ee-uh). Twinspur.
Dicentra (dy-SEN-truh). Bleeding heart, Dutchman's-breeches.
Dichondra (dy-KON-druh). Creeping tropical vines; morning-glory family.
Dichotoma (dy-KOT-o-muh). Forked in pairs.
Dicksonia (dik-SOH-nee-uh). Greenhouse tree ferns.
Dicotyledon (dy-kot-ih-LEE-don). Plants with two cotyledons.
Dictamnus (dik-TAM-nus). Gas plant.
Didymus (DID-ih-mus). In pairs, as stamens.
Dieffenbachia (deef-en-BAH-kee-uh). Greenhouse foliage plants.
Diervilla (dy-ur-VIL-uh). Bush honeysuckle.
Diffusus (dih-FYOO-sus). Spreading.
Dierama (dee-uh-RAH-muh). Wand flower.
Digitalis (dih-jih-TAY-liss). Foxglove.
Digitaria (dih-jih-TAY-ree-uh). Crab grass.
Digitate (DIH-jih-tayt). Fingered in a whorl, hand-like.
Dilatatus (dy-luh-TAY-tus). Dilated, expanded.
Dimerous (DIM-ur-us). All parts in twos.
Dimorphotheca (dy-mor-fo-THEEK-uh). Double-fruited.
Dimorphous (dy-MOR-fus). Occurring in two forms.
Dioecious (dy-EE-shus). Male and female flowers on separate plants.
Dionaea (dy-o-NEE-uh). Venus's-flytrap.
Dioon (dy-OH-on). With pairs of seeds.
Dioscorea (dy-oss-KO-ree uh). Yam, cinnamon vine.
Diosma (dy-OZ-muh; dy-OSS-muh). Tender heath-like shrubs.
Diospyros (dy-OSS-pih-ros). Persimmon, ebony.

Diotis (dy-O-tiss). Cotton weed.
Dipelta (dy-PEL-tuh). Hardy flowering shrubs.
Dipladenia (dip-luh-DEE-nee-uh). Showy tropical woody vines.
Diplazium (dip-LAY-zee-um). With a double indusium.
Diplotaxis (dip-lo-TAK-siss). Rocket.
Dipsacus (DIP-suh-kus). Teasel.
Dipterus (DIP-tur-us). Two-winged.
Dirca (DUR-kuh). Leatherwood.
Discolor (DISS-kuh-lur). Of two or different colors.
Disporopsis (diss-por-OP-siss). With paired seeds.
Disporum (dy-SPOH-rum). Fairy bells.
Distichus (DISS-tih-kus). In two vertical ranks.
Distictis (diss-TIK-tiss). Double-spotted.
Diurnus (dy-UR-nus). Day-flowering.
Divaricate (dy-VAR-ih-kayt). Widely divergent, spreading.
Divergens (dy-VUR-jenz). Wide-spreading.
Diversiflorus (dih-ver-sih-FLO-rus). Diversely flowered.
Diversifolia (dy-ver-sih-FO-lee-uh). Leaves variable.
Dizygotheca (dy-zy-go-THEEK-uh). False aralia.
Dodecatheon (doh-deh-KATH-ee-on). Shooting star, N. American wild flowers.
Dodonaea (doh-don-EE-uh). From proper name.
Dolichos (DOL-ih-kos). Tropical vines; hyacinth bean.
Dombeya (dom-BEE-yuh; DOM-bee-yuh). Showy African trees, shrubs.
Donax (DOH-naks). Giant reeds.
Doodia (DOO-dee-uh). Hacksaw fern.
Doronicum (doh-RON-ih-kum). Leopard's bane.
Dorotheanthus (dor-uh-thee-AN-thus). From proper name.
Doryanthes (dor-ee-AN-theez). Plants like agaves; spear lily.
Dorycnium (doh-RIK-nee-um). Clover-like herbs.
Doryopteris (dor-ree-OP-tur-iss). Small tropical American ferns.
Douglasia (dug-LASS-ee-uh). European alpine plants.
Dovyalis (doh-vee-AY-liss). Subtropical fruit plants.
Downingia (dow-NIN-jee-uh). Hardy annuals.

Doxantha (doks-AN-thuh). Tropical woody vines; cat's-claw.
Draba (DRAY-buh). Rock garden crucifers.
Dracaena (druh-SEE-nuh). Handsome foliage plants.
Draco (DRAY-ko). Pertaining to a dragon.
Dracocephalum (dray-ko-SEF-uh-lum). Dragonhead.
Dracunculus (dray-KUN-kyoo-lus). Curious aroids; green dragon.
Drimys (DRIH-muss). Sharp, stinging (bark taste).
Drosanthemum (draw-SAN-theh-mum). Flower dew.
Drosera (DROS-ur-uh). Insectivorous bog herbs; sundew.
Drupe (dryoop). Stone fruit, like cherry, peach, plum.
Dryandra (dry-AN-druh). Australian shrubs.
Dryas (DRY-as). Evergreen rockery creepers.
Dryopteris (dry-OP-teer-iss). Hardy wood ferns.
Duchesnea (dyoo-KES-nee-uh). Mock strawberry.
Dudleya (DUD-lee-uh). Succulent West Coast perennials.
Dulcamara (dul-kuh-MAY-ruh). Climbing nightshade.
Dulcis (DUL-siss). Sweet.
Dumosus (dyoo-MO-sus). Bushy, shrubby.
Duranta (dyoo-RAN-tuh). Tropical lilac, pigeonberry.
Durio (DYOO-ree-o). Malayan trees; durian.
Dyckia (DY-kee-uh). From proper name.

E

Eburneus (ee-BUR-nee-us). Ivory.
Ecballium (ek-BAW-lee-um). Squirting cucumber.
Eccremocarpus (ek-kreh-mo-KAHR-pus). With hanging fruit.
Echeveria (ek-eh-VEE-ree-uh; ech-eh-VEE-ree-uh). Stemless succulents.
Echinacea (ek-ih-NAY-see-uh). Purple coneflower.
Echinatus (ek-ih-NAY-tus). Prickly, bristly.
Echinocactus (ee-ky-no-KAK-tus). Cylindrical or globular cacti.
Echinocereus (ee-ky-no-SEE-ree-us; ek-ih-no-SEE-ree-us). Low clump cacti.
Echinocystis (ee-ky-no-SISS-tiss). Wild cucumber.

Echinopanax (ee-ky-no-PAY-naks; ek-ih-no-PAY-naks). Devil's-club.
Echinops (EK-ih-nops). Globe thistles.
Echinopsis (ek-ih-NOP-siss). Sea urchin cactus.
Echium (EK-ee-um). Viper's bugloss, blueweed.
Edulis (ED-yoo-liss; eh-DYOO-liss). Edible.
Ehretia (eh-REE-tee-uh; eh-REE-she-uh). Semitropical trees or shrubs.
Eichhornia (eyek-HOR-nee-uh). Water hyacinth.
Elaeagnus (el-ee-AG-nus; ee-lee-AG-nus). Russian olive and others.
Elaphoglossum (el-uh-fo-GLOSS-um). Elephant-ear fern.
Elasticus (ee-LASS-tih-kus). Rubber-producing, elastic.
Elatus (ee-LAY-tus). Tall.
Elegans (EL-eh-ganz). Elegant, beautiful.
Elegantissimus (el-eh-gan-TISS-ih-mus). Most elegant or beautiful.
Elettaria (el-eh-TAY-ree-uh). Cardamom.
Eleusine (el-yoo-SY-nee). Wire grass, African millet.
Elliottianum (el-ee-ot-ee-AY-num). From proper name.
Ellipticus (el-LIP-tik-us). Elliptical.
Elodea (el-o-DEE-uh; ee-LO-dee-uh). Common aquarium plants.
Elongatus (ee-lon-GAY-tus). Elongated, lengthened.
Elsholtzia (el-SHOLT-see-uh). Mintshrub, heathermint.
Elymus (EL-ih-mus). Wild rye, lyme grass.
Emarginata (ee-mahr-jih-NAY-tuh). Shallow notch at apex.
Emilia (ee-MIL-ee-uh). Tassel flower.
Empetrum (em-PE-trum). Evergreen, prostrate or mat-forming shrubs; N. America and Andes.
Encelia (en-SEE-lee-uh). Shrubbery composites; brittlebush.
Endocarp (EN-doh-karp). Inner layer of pericarp.
Endogen (EN-doh-jen). A monocotyledon.
Enkianthus (en-kee-AN-thus). Shrubs of heath family.
Entelea (en-TEE-lee-uh; en-teh-LEE-uh). Corkwood.
Entire (en-TIRE). Without toothing or division.

Entomology (en-toh-MAH-lo-jee). Study of insects.
Epacris (EP-uh-kriss). Heath-like evergreen shrubs.
Ephedra (eh-FEE-druh). Leafless desert shrubs; joint fir.
Ephemeral (eh-FEHM-ur-al). Lasting only one day.
Epidendrum (ep-ih-DEN-drum). Tropical American orchids.
Epigaea (ep-ih-JEE-uh). Trailing arbutus.
Epigynous (ee-PIH-jin-us). Joined to the ovary.
Epilobium (ep-ih-LO-bee-um). Willow herb, fireweed.
Epimedium (ep-ih-MEE-dee-um). Woody rockery perennials.
Epipactis (ep-ih-PAK-tiss). Hardy woodland orchids.
Epiphyllum (ep-ih-FIL-um). Epiphytic spineless cacti.
Epiphyte (EP-ih-fyte). Air plant; not parasitic.
Epipremnum (ep-ee-PREM-noom). On or beside a tree trunk.
Episcia (eh-PISS-ee-uh). Showy drooping greenhouse plant.
Equestris (ee-KWESS-triss). Pertaining to the horse.
Equisetum (ek-wih-SEE-tum). Horsetail.
Eragrostis (er-uh-GROS-tiss). Love grass.
Eranthemum (ee-RAN-thee-mum). Attractive greenhouse shrub.
Eranthis (ee-RAN-thiss). Winter aconite.
Erectus (ee-REK-tus). Upright, erect.
Eremophila (ayr-eh-MOFF-ih-luh). Solitary love.
Eremurus (er-eh-MYOO-rus). Desert candle, foxtail lily.
Eria (EE-ree-uh). Asian epiphytic orchids.
Erianthus (er-ee-AN-thus). Plume grasses; Ravenna grass.
Erica (ee-RY-kuh; ER-ih-kuh as common name). True heath.
Ericaceae (er-ih-KAY-see-ee). Heath family.
Ericaceous (er-ih-KAY-shus). Pertaining to members of heath family.
Ericameria (er-ih-kuh-MEE-ree-uh). Mock heath.
Ericoides (er-ih-KOY-deez). Resembling a heath.
Erigeron (ee-RIH-jur-on). Aster-like plants; daisy fleabane.
Erigenia (er-ih-JEE-nee-uh). Harbinger-of-spring.
Erinus (eh-RY-nus). Tiny, tufted hardy perennials.
Eriobotrya (ee-ree-oh-BOT-ree-uh; eh-ree-oh-BOT-ree-uh). Loquat.

Eriocephalus (ee-ree-oh-SEF-uh-lus). Aromatic African shrubs.
Eriogonum (ee-ree-OG-o-num). Wild buckwheat.
Eriophorum (ee-ree-OFF-oh-rum; er-ee-OFF-oh-rum). Bog sedge; cotton grass.
Eriophorus (er-ee-OFF-o-rus). Wool-bearing, woolly.
Eriophyllum (ee-ree-o-FIL-um; er-ee-o-FIL-um). Woolly sunflower.
Eriostemon (er-ee-oh-STEE-mon). Australian evergreen shrubs.
Eritrichium (er-ih-TRIK-ee-um). Woolly alpine plants.
Erlangea (ur-LAN-jee-uh). Tall ageratum-like plants.
Erodium (ee-ROW-dee-um). Heron's-bill, alfilarias.
Erosus (ee-ROW-sus). Jagged, as if gnawed.
Erraticus (er-RAT-ih-kus). Erratic, unusual, sporadic.
Erubescens (er-yoo-BESS-senz). Blushing, red or reddish.
Ervatamia (er-vuh-TAY-mee-uh). Crape jasmine.
Eryngium (eh-RIN-jee-um). Sea holly, button snakeroot.
Erysimum (eh-RISS-ee-mum). Alpine wallflower.
Erythea (er-ih-THEE-uh). Mexican fan palms.
Erythrina (er-ih-THRY-nuh). Coral tree, bucare.
Erythronium (er-ih-THROW-nee-um). Dogtooth violet, trout lily.
Erythroxylon (er-ih-THROK-sih-lon). Coca, cocaine plants.
Escallonia (ess-kah-LOH-nee-uh). Handsome southern shrubs.
Eschscholtzia (eh-SHOLT-see-uh). California poppy.
Escobaria (ess-ko-BAY-ree-uh). Mexican globular cacti.
Escontria (ess-KON-tree-uh). Tree-like Mexican cactus.
Esculentus (ess-kyoo-LEN-tus). Esculent, edible.
Eucalyptus (yoo-kuh-LIP-tus). Australian evergreen trees.
Eucharidium (yoo-kuh-RID-ee-um). Fairy fan, red ribbons.
Eucharis (YOO-kuh-riss). Amazon lily.
Eucomis (YOO-ko-miss). Pineapple flower.
Eucommia (yoo-KOM-ee-uh). Elm-like rubber-producing trees.
Eucryphia (yoo-KREE-fee-uh). Well-covered (flower bud).
Eugenia (yoo-JEE-nee-uh). Tropical aromatic trees, shrubs.
Eulalia (yoo-LAY-lee-uh). Tall grasses; now *Miscanthus*.

Euonymus (yoo-ON-ih-mus). Hardy shrubs and vines, evergreen and deciduous.
Eupatorium (yoo-puh-TOH-ree-um). Hardy ageratum.
Euphorbia (yoo-FOR-bee-uh). Poinsettia, crown of thorns, spurge.
Euphoria (yoo-FO-ree-uh). Subtropical fruit trees.
Euptelea (yoo-TEE-lee-uh). Small, showy Asiatic trees.
Europaeus (yoo-ro-PEE-us). European.
Eurya (YOO-ree-uh). Evergreen foliage shrubs.
Euryops (YOO-ree-ops). Big eyes (flowers).
Eustoma (yoo-STO-muh). Prairie gentian, bluebell.
Euterpe (yoo-TUR-pee). Feather palm; assai palm.
Evodia (ee-VO-dee-uh). Sumac-like shrubs and trees.
Evolvulus (ee-VOL-vyoo-lus). Unravel.
Evonymus (ev-ON-ih-mus). Synonym of euonymus.
Exacum (EK-suh-kum). Summer-flowering pot plants.
Exaltata (eks-awl-TAY-tuh). Extremely tall.
Excelsus (ek-SEL-sus). Tall.
Excisus (ek-SY-sus). Cut away.
Exiguus (eks-IG-yoo-us). Small, meager.
Eximius (eks-IM-ee-us). Distinguished, unusual.
Exochorda (ek-so-KOR-duh). Pearlbush.
Exotic (eg-ZOT-ik). Not native, foreign.
Exserted (eks-SUR-ted). Projecting.
Extrorse (eks-TRORS). Facing outward.

F

Fabaceus (fay-BAY-see-us; fah-BAY-see-us). Bean-like.
Fabiana (fay-bee-AY-nuh). False heath.
Fagopyrum (fag-o-PY-rum). Buckwheat.
Fagus (FAY-gus). Beech.
Falcatus (fal-KAY-tus). Like a sickle.
Falciformis (fal-sih-FOR-miss). Sickle-shaped.
Fallax (FAL-aks). Deceptive.
Farinaceous (fayr-ih-NAY-shus). Starch-like, mealy.

Farleyense (fahr-lee-ENSS). After Farley Hill garden, Barbados.
Fasciate (FASH-ee-ayt). Broad, flattened stem.
Fascicle (FAS-ih-kul). Bundle, compact cluster.
Fastigiatus (fass-tih-jee-AY-tus). Erect, sides nearly parallel.
Fastuosus (fast-yoo-O-sus). Stately, proud, bountiful.
Fatshedera (fats-HED-eh-ruh). Bigeneric hybrid; English ivy and *Fatsia japonica.*
Fatsia (FATS-see-uh). Tender evergreen Japanese shrubs.
Feijoa (fay-JO-uh; fee-JO-uh). Subtropical fruit; pineapple guava.
Felicia (feh-LISS-ee-uh). Blue daisy.
Ferocactus (fee-roh-KAK-tus). Fiercely spiny globular cacti.
Ferruginous (feh-ROO-jih-nus). Rust color.
Fertilis (FUR-til-iss). Productive, fruitful.
Ferula (FER-yoo-luh). Giant fennel.
Festivus (fess-TY-vus). Festive, gay, bright.
Festuca (fess-TYOO-kuh). Fescue grass.
Ficus (FY-kus). Fig, rubber plant.
Filamentosa (fil-uh-men-TOH-suh). Thread-like.
Filicales (fil-ih-KAY-leez). Order of all true ferns.
Filicifolia (fih-liss-ih-FO-lee-uh). Fern-leaved.
Filiciform (fih-LISS-ih-form). Fern-shaped.
Filiferus (fih-LIF-ur-us). Bearing threads.
Filifolius (fil-ih-FO-lee-us). With thread-like leaves.
Filiform (FIL-ih-form; FY-lih-form). Thread shaped.
Filipendula (fil-ih-PEN-dyoo-luh). Meadowsweet, dropwort.
Filix-femina (fy-liks-FEM-ih-nuh). Lady fern.
Filix-mas (fy-liks-MAS). Male fern.
Fimbriata (fim-bree-AY-tuh). Fringed.
Firmiana (fur-mee-AY-nuh). Chinese parasol tree, Phoenix tree.
Fistulosus (fiss-tyoo-LO-sus). Hollow, cylindrical.
Fittonia (fih-TOH-nee-uh). Greenhouse foliage plants.
Fitzroya (fitz-ROY-uh). Chilean evergreen tree; pine family.
Flabelliformis (fluh-bel-ih-FOR-miss). Fan-shaped.
Flaccida (FLAK-sid-uh). Flabby.
Flagellaris (flaj-eh-LAY-riss). Whip-like.

Flagelliformis (fluh-jel-ih-FOR-miss). Whip-shaped.
Flagellum (fluh-JEL-um). Whip-like organ.
Flammeus (FLAM-ee-us). Flame-colored.
Flavescens (flay-VESS-enz). Yellowish.
Flavus (FLAY-vus). Yellow.
Flemingia (fleh-MIN-jee-uh). Southern flowering shrub.
Flexilis (FLEKS-ih-liss). Flexible, pliant.
Flexuosus (fleks-yoo-O-sus). Zigzag, tortuous.
Floccosus (flok-KO-sus). Woolly.
Flore-pleno (flo-reh-PLEE-no). With double flowers.
Floribundus (flor-ih-BUN-dus). Free flowering.
Floriculture (FLO-rih-kul-tur; FLOR-ih-kul-tur). Culture of flowers, ornamental plants.
Floridus (FLO-rih-dus). Flowering freely.
Floriferous (flo-RIF-ur-us). Blooming freely.
Foemina (FEEM-ih-nuh; FEH-mih-nuh as common name). Feminine.
Foeniculum (fee-NIK-yoo-lum). Fennel.
Foetidus (FET-ih-dus; FEE-tid-us). Fetid, bad-smelling.
Foliosus (fo-lee-O-sus). Full of leaves.
Fontanesia (fon-tuh-NEE-zee-uh). Privet-like shrubs.
Fontinalis (fon-tih-NAY-liss). Pertaining to a spring.
Forestiera (for-ess-tih-EE-ruh). Swamp privet.
Formosanus (for-mo-SAY-nus). Of Formosa (Portuguese name for Taiwan).
Formosissimus (for-mo-SISS-ih-mus). Very beautiful.
Formosum (for-MO-sum). Handsome, beautiful.
Forsterianum (fors-ter-ee-AY-num). From proper name.
Forsythia (for-SITH-ee-uh; for-SY-thee-uh). Golden bells.
Fortunei (for-TYOO-nee-eye). From proper name.
Fortunella (for-tyoo-NEL-uh). Kumquat.
Fothergilla (foth-ur-GIL-uh). Springscent, woody plants.
Fouquieria (foo-kih-EE-ree-uh). Ocotillo, Jacob's-staff.
Fragaria (fruh-GAY-ree-uh). Strawberry.
Fragilis (FRAJ-ih-liss). Fragile, brittle.

Fragrans (FRAY-granz). Fragrant.
Fragrantissima (fray-gran-TISS-ih-muh). Most fragrant.
Francoa (frang-KO-uh). Perennial herbs of Chile.
Frangula (FRANG-yoo-luh; FRAN-gyoo-luh). Alder buckthorn.
Franklinia (frank-LIN-ee-uh). Franklin tree.
Frasera (FRAY-zur-uh). American columbo.
Fraxineus (frak-SIN-ee-us). Resembling the ash.
Fraxinus (frak-SIH-nus). Ash.
Freesia (FREE-zhee-uh; FREE-see-uh). Winter-flowering bulbs.
Fremontia (free-MON-tee-uh). Flannelbush, leatherwood.
Fritillaria (frit-ih-LAY-ree-uh). Checkered lily, crown imperial.
Frigidus (FRIJ-ih-dus). Cold, of cold regions.
Frondosus (fron-DOH-sus). Leafy.
Fructescent (fruk-TESS-ent). Fruitful.
Frutescens (froo-TESS-enz). Shrubby, bushy.
Fruticosus (froo-tih-KO-sus). Shrubby.
Fuchsia (FYOOK-see-uh; FYOO-shuh; FYOO-she-uh as common name). Lady's-eardrops.
Fuchsioides (fyook-zee-OY-deez). Fuchsia-like.
Fugacious (fyoo-GAY-shus). Withering quickly.
Fulgens (FUL-genz). Shining, glistening.
Fulgidus (FUL-jih-dus). Shining.
Fulvus (FUL-vus). Tawny, orange-gray-yellow.
Fumaria (fyoo-MAY-ree-uh). Fumitory.
Fumosus (fyoo-MO-sus). Smoky.
Fungi (FUN-jeye). Flowerless plants lacking chlorophyll.
Fungicide (FUN-jih-side). Substance that prevents fungi from developing.
Fungous (FUNG-gus). Adjective form of fungus.
Funkia (FUNG-kee-uh). Plantain lily; now *Hosta*.
Furcatus (fur-KAY-tus). Forked.
Furcraea (fur-KREE-uh). Agave-like plants; giant lily.
Fuscous (FUS-kus). Grayish brown.
Fusiform (FYOO-zih-form; fyoo-sih-form). Spindle-shaped.

G

Gaillardia (gay-LAHR-dee-uh). Blanketflower.
Galacifolia (gay-lass-ih-FO-lee-uh). Leaves like galax.
Galanthus (guh-LAN-thus). Snowdrops.
Galax (GAY-laks). Evergreen mountain herbs.
Gale (GAYL). Myrtle bush.
Galeate (GAY-lee-ayt). Helmet-shaped.
Galega (guh-LEE-guh). Goat's-rue.
Galium (GAY-lee-um). Bedstraw.
Gallicus (GAL-lih-kus). Of Gaul (France) or pertaining to a cock.
Galtonia (gawl-TOH-nee-uh). Summer hyacinth.
Gamopetalous (gam-o-PET-al-us). Petals united.
Gandavensis (gan-duh-VEN-siss). Of Ghent, Belgium.
Gardenia (gahr-DEE-nee-uh). Evergreen white-flowered shrubs.
Garrya (GAR-ee-uh). Silk-tassel tree, bear brush.
Gaskelliana (gas-kel-ee-AY-nuh). From proper name.
Gasteria (gas-TEE-ree-uh). Aloe-like desert plants.
Gaultheria (gawl-THEE-ree-uh). Wintergreen; checkerberry.
Gaura (GAW-ruh). Proud, majestic.
Gaussia (GAWS-ee-uh). Cuban feather palm.
Gaylussacia (gay-lyoo-SAY-she-uh). Huckleberry.
Gazania (gay-ZAY-nee-uh). Showy African garden herbs.
Gelsemium (jel-SEE-mee-um). Carolina or yellow jessamine.
Geminiflorus (jem-ih-nih-FLO-rus). Flowers in pairs.
Gemmatus (jem-MAY-tus). Bearing buds.
Genera (JEN-eh-rah). Plural of genus.
Generic (jeh-NEH-rik). Pertaining to genus.
Geniculatus (jen-ik-yoo-LAY-tus). Bent, jointed.
Geniostoma (jen-ee-OSS-toh-muh). Shrubs of south Pacific.
Genipa (GEN-ih-puh; jee-NY-puh). Tropical fruit trees; genipap.
Genista (jeh-NISS-tuh). Spiny shrubs of pea family; broom.
Gentian (JEN-shan). Blue-flowered garden plants.
Gentiana (jen-she-AY-nuh; jen-she-AN-uh). Gentian.

Genus (JEE-nus). A group of related species.
Geonoma (jee-ON-o-muh). Tropical American feather palms.
Geranium (jeh-RAY-nee-um). Wild or hardy geranium, crane's-bill.
Gerbera (jur-BEE-ruh; gur-BEE-ruh). African or Transvaal daisy.
Gesneria (jess-NEE-ree-uh). Showy tropical herbs.
Geum (JEE-um). Avens; popular border perennials.
Gibbaeum (jih-BE-um). Hump-shaped (some species).
Giganteum (jy-gan-TEE-um; jih-GAN-tee-um). Gigantic.
Gigas (GY-gas). Of giants, immense.
Gilia (GIL-ee-uh; JIL-ee-uh). Showy garden herbs.
Ginkgo (GINGK-go; JINGK-go). Maidenhair tree.
Ginnala (gih-NAY-luh). Asiatic vernacular name, specific name.
Glabra (GLAY-bruh). Smooth, without hairs.
Glabrous (GLAY-brus). Smooth, not hairy.
Glacialis (glay-she-AY-lus; glay-see-AY-liss). Growing on or near a glacier.
Gladioli (glad-ee-O-lye; gluh-DY-o-lye). Plural of gladiolus.
Gladiolus (glad-ee-O-lus; gluh-DY-o-lus). Popular carnous plants.
Gladioluses (glad-ee-O-lus-ez; glad-ee-O-lus-iz). Plural form recognized.
Glandulosus (glan-dyoo-LO-sus). Bearing glands.
Glauca (GLAW-kuh). With white or gray bloom, as on blue spruce.
Glaucescens (glaw-SESS-enz). Becoming or almost glaucous.
Glaucidium (glaw-SID-ee-um). Rock garden herbs.
Glaucium (GLAW-see-um). Horn poppy.
Glaucophylla (glaw-ko-FIL-uh). Foliage glaucous.
Glaucous (GLAW-kus). Whitened with minute powder.
Gleditsia (gleh-DIT-see-uh; glee-DIT-see-uh). Honey locust.
Globosa (glo-BO-suh). Globe-shaped.
Globularia (glob-yoo-LAY-ree-uh). Globe daisy.
Globuliferus (glob-yoo-LIF-ur-us). Globule or globe-bearing.
Glomerate (GLOM-ur-ayt). Compactly clustered.
Gloriosa (glo-ree-O-suh). Glory or climbing lily.

Glottiphyllum (glah-tih-FILL-um). With tongue-like leaves.
Gloxinia (glok-SIN-ee-uh). Showy florists' plants.
Glutinosus (gloo-tih-NO-sus). Glutinous, sticky.
Gnaphalium (nuh-FAY-lee-um). Woolly herbs.
Godetia (go-DEE-she-uh). Showy garden annuals.
Gomesa (go-MEE-zhuh). Brazilian orchids.
Gomphrena (gom-FREE-nuh). Globe amaranth.
Goniolimon (gaw-nee-oh-LEE-mon). Perennial herbs.
Goodia (go-O-dee-uh). Australian shrubs.
Goodyera (good-YEER-uh). Rattlesnake plantain; from proper name.
Gordonia (gor-DOH-nee-uh). Franklin tree—see *Franklinia*.
Gormania (gor-MAY-nee-uh). Succulents like sedums.
Gossypium (gah-SEE-pee-um). Cotton.
Grabowskia (gruh-BOH-skee-uh). Tender spiny shrubs.
Gracilis (GRASS-ih-liss). Graceful, slender.
Graminifolious (gram-ih-nih-FO-lee-us). Grass-like foliage.
Gramineae (gruh-MIN-ee-ee). Grass family.
Grammatophyllum (gram-uh-toh-FIL-um). Malayan orchids.
Granatum (gruh-NAY-tum). Early name of pomegranate.
Grandiflorus (gran-dih-FLOR-us). Large-flowered.
Grandis (GRAN-diss). Large, showy.
Graniticus (gruh-NIT-ih-kus). Granite-loving.
Graptopetalum (grap-toh-PET-uh-lum). Marked (variegated) leaves.
Gratianopolitanus (grah-tee-AH-no-pol-ih-TAY-nus). Of Grenoble, France.
Gratissimus (gruh-TISS-ih-mus). Very pleasing or agreeable.
Graveolens (gruh-VEE-o-lenz). Heavy-scented.
Grevillea (gre-VIL-ee-uh). Silk oak.
Grewia (GROO-ee-uh). Tropical shrubs, trees.
Greyia (GRAY-ee-eye). Small African trees.
Grindelia (grin-DEE-lee-uh). Gummy herbs; gum plant.
Griselinia (grih-seh-LIN-ee-uh). Tender evergreen trees, shrubs.
Guava (GWAH-vuh). Common name for *Psidium*.

Guianensis (gee-an-EN-siss). Of Guiana.
Gunnera (GUN-ur-uh; gu-NEE-ruh). Showy foliage plants.
Guttatus (guh-TAY-tus). Spotted, speckled.
Guzmania (guz-MAY-nee-uh). Mostly epiphytic bromeliads.
Gymnocalycium (jim-no-kuh-LEE-see-um). With naked buds.
Gymnocarpa (jim-no-KAHR-puh). With naked fruit.
Gymnocarpium (jim-no-KAHR-pee-um). With naked fruit.
Gymnocladus (jim-NOK-luh-dus). Kentucky coffee tree.
Gymnosperm (JIM-no-spurm). Plant bearing naked seeds—conifer, cycad.
Gynerium (jy-NEE-ree-um). Tall perennial grasses.
Gynura (jy-NYOO-ruh; jih-NYOO-ruh). Velvet plant.
Gypsophila (jip-SOF-ih-luh). Baby's-breath.

H

Haageanum (hah-gee-AY-num). From proper name.
Habenaria (hab-eh-NAY-ree-uh). Fringed orchid.
Haberlea (ha-BUR-lee-uh). Tufted rock garden herbs.
Haemanthus (hee-MAN-thus). Blood lily.
Hakea (HAH-kee-uh). Evergreen shrubs; cushionflower.
Halesia (ha-LEE-zhee-uh; ha-LEE-see-uh). Silver bell.
Halimifolia (ha-lih-mih-FO-lee-uh). Halimus-leaved.
Halimodendron (ha-lih-mo-DEN-dron). Salt tree.
Halliana (hawl-ee-AY-nuh; hal-ee-AY-nuh). From proper name, Hall.
Hamamelis (ham-uh-MEE-liss). Witch hazel.
Hamelia (huh-MEE-lee-uh). Tropical shrubs; scarlet bush.
Hansoni (han-SO-nee). From proper name.
Harbouria (hahr-BOO-ree-uh). Feathery rock garden herb.
Hardenbergia (hahr-den-BUR-jee-uh). Tender woody vines.
Harrisi (hay-RISS-eye). From proper name.
Harrisia (ha-RISS-ee-uh). Vine-like cacti.
Harrisoniae (hayr-ih-SO-nee-ee). From proper name.
Hartwegia (hart-WEE-jee-uh). Tropical orchids.

Hastata (has-TAY-tuh). Like arrowhead.
Hatiora (hat-ee-OR-uh). Cacti genus; from proper name.
Haworthia (huh-WUR-thee-uh; huh-WOR-thee-uh). Rosetted succulents.
Hebe (HEE-bee). Veronica-like shrubs.
Hedeoma (hee-dee-O-muh). American pennyroyal.
Hedera (HED-ur-uh). Generic name of English ivy.
Hederaceus (hed-ur-AY-see-us). Of or like ivy.
Hedychium (hee-DIK-ee-um). Ginger or butterfly lily.
Hedyscepe (hed-ih-SEE-pee). Umbrella palm.
Helenium (hel-EE-nee-um). Sneezeweed.
Heliamphora (hel-ee-AM-for-uh). Pitcher plant.
Helianthemum (hee-lee-AN-thee-mum). Sun rose, rockrose.
Helianthus (hee-lee-AN-thus). Sunflower.
Helichrysum (hel-ih-KRY-sum). Strawflower, everlasting.
Heliconia (hel-ih-KO-nee-uh). Wild plantain, balisier.
Heliocereus (hee-lee-O-SEE-ree-us). Sprawling Mexican cacti.
Helictotrichon (hee-lik-toh-TRY-kon). Oatgrass.
Heliophila (hee-lee-OFF-ih-luh). Cape stock.
Heliopsis (hee-lee-OP-siss). Coarse sunflower-like herbs.
Heliotropium (hee-lee-o-TRO-pee-um). Heliotrope.
Helipterum (hee-LIP-tur-um). Swan River everlasting, immortelles.
Helix (HEL-iks). Species name of English ivy (*Hedera*).
Helleborus (hel-LEB-o-rus). Christmas rose.
Helonias (hee-LO-nee-us). Swamp pink.
Helxine (hel-ZY-nee). Baby tears.
Hemerocallis (hem-ur-o-KAL-iss). Day lily.
Hemionitis (hem-ee-o-NY-tiss). Strawberry or ivy fern.
Hemiptelea (hem-ip-TEL-ee-uh). Half-winged (fruit).
Hendersoni (hen-dur-SO-nee). From proper name.
Henryi (HEN-ree-eye). From proper name.
Hepatica (hee-PAT-ih-kuh). Spring-blooming wild flowers.
Heracleum (her-uh-KLEE-um). Cow parsnip.

Herbaceous (hur-BAY-shus). Lacking persistent stem above ground.

Herniaria (hur-nee-AY-ree-uh). Carpet bedding herbs.

Hesperis (HES-pur-iss). Rocket, damewort, sweet rocket.

Heterocentron (het-ur-oh-SEN-tron). With variable-sized spurs.

Heterogamous (het-ur-OG-uh-mus). Bearing two kinds of flowers.

Heteromeles (het-ur-o-MEE-leez). Toyon, California holly.

Heterophyllus (het-ur-o-FIL-us). Leaves vary on same plant.

Heterospathe (het-ur-OS-puh-thee). Sagisi palm.

Heuchera (HYOO-kur-uh). Coral-bells, alumroot.

Heucherella (hyoo-ker-EL-uh). Hybrid of *Heuchera* and *Tiarella*.

Hevea (HEE-vee-uh). Para rubber tree.

Hibbertia (hih-BUR-tee-uh). Showy woody vines, shrubs.

Hibernalis (hy-bur-NAY-liss). Pertaining to winter.

Hibernicus (hy-BUR-nee-kus). Of Ireland.

Hibiscus (hy-BISS-kus). Mallow, okra, roselle.

Hieracium (hy-ur-AY-she-um; hy-ur-AY-see-um). Hawkweed.

Hierochloe (hy-ur-OK-lo-ee). Holy, vanilla or Seneca grass.

Himalaicus (him-uh-LAY-ih-kus). Himalayan.

Hippeastrum (hip-ee-ASS-trum). Amaryllis-like plants.

Hippocastanum (hip-o-kass-TAY-num). Specific name of horse chestnut, *Aesculus*.

Hippophae (hy-POF-ay-ee). Sea buckthorn.

Hirsuta (her-SYOOT-uh). With coarse or stiff hairs.

Hirta (HUR-tuh). Hairy.

Hispanicus (hiss-PAN-ih-kus). Of Spain.

Hispida (HISS-pih-duh). With bristly hairs.

Hoffmannia (hoff-MAN-ee-uh). Greenhouse foliage plants.

Hoheria (ho-HEE-ree-uh). Ribbonwood, lacebark.

Holcus (HOL-kus). Johnson grass, sorghum, velvet grass.

Holodiscus (hol-o-DISS-kus). Rock spirea.

Homalocephala (ho-mal-o-SEFF-uh-luh). A Southwestern cactus.

Homogamous (ho-MOG-uh-mus). Bearing one kind of flower.

Hordeum (HOR-dee-um). Barley, squirreltail grass.

Horizontalis (hor-ih-zon-TAY-liss). Horizontal.
Horminum (hor-MY-num). Rock garden labiate.
Horridus (HOR-ih-dus). Prickly.
Hortensis (hor-TEN-siss). Of the garden.
Horticulture (HOR-tih-kul-tyur). Art and science of growing flowers, fruits and vegetables.
Hosackia (ho-SAK-ee-uh). Witch's-teeth.
Hosta (HOSS-tuh). Plantain lily.
Houstonia (hoos-TOH-nee-uh). Bluet.
Hovea (HO-vee-uh). Australian shrubs.
Hovenia (ho-VEE-nee-uh). Japanese raisin tree.
Howea (how-ee-uh). Florists' feather palms.
Hoya (HOY-uh). Tropical vines; wax plant.
Humata (hyoo-MAY-tuh). Bear's-foot ferns.
Humilis (HYOO-mih-liss). Low-growing, dwarf.
Humulus (HYOO-myoo-lus). Hops.
Hunnemannia (hun-neh-MAN-nee-uh). Mexican tulip poppy.
Huntleya (HUNT-lee-uh). Tender epiphytic orchids.
Hupehensis (hyoo-peh-EN-siss). From Hupeh, China.
Hutchinsia (hut-CHIN-see-uh). Tiny alpine herbs.
Hyacinth (HY-uh-sinth). Common bulbous plants.
Hyacinthus (hy-uh-SIN-thus). Hyacinth.
Hybridus (HY-brih-dus). Hybrid, mixed, mongrel.
Hydrangea (hy-DRAN-jee-uh). Hardy shrubs, vines; florists' plants.
Hydrastis (hy-DRAS-tiss). Goldenseal, orangeroot.
Hydriastele (hy-dree-as-TEE-lee). Australian feather palm.
Hydrocotyle (hy-droh-KOT-ul-ee). Water cup (leaf form).
Hydrophyllum (hy-dro-FIL-um). Waterleaf.
Hydrosme (hy-DROSS-mee). Devil's tongue, snake palm.
Hyemalis (hy-eh-MAY-liss). Of winter.
Hylocereus (hy-lo-SEE-ree-us). A night-blooming cereus.
Hymenanthera (hy-meh-NAN-thur-uh). Stiff tender shrubs.
Hymenocallis (hy-meh-no-KAL-iss). Peruvian lily.
Hymenophyllum (hy-meh-no-FIL-um). Tropical ferns.

Hymenosporum (hy-meh-NOS-po-rum). Australian shrub.
Hyophorbe (hy-o-FOR-bee). Pignut, bottle, spindle palm.
Hyoscyamus (hy-o-SY-uh-mus). Henbane.
Hypericoides (hy-pur-ih-KOY-deez). Hypericum-like.
Hypericum (hy-PEH-rih-kum). St.-John's-wort.
Hyphaene (hy-FEE-nee). African fan palm; doum palm.
Hypochoeris (hy-po-KEE-riss). Cat's-ear.
Hypolepis (hy-POLL-eh-piss). Tropical ferns.
Hypoleucus (hy-po-LOO-kus). White underneath.
Hypoxis (hy-POK-siss; hih-POK-siss). Star grass.
Hyssop (HISS-op; HIZZ-op). Aromatic herb of mint family.
Hyssopifolia (hiss-o-pih-FO-lee-uh). With hyssop-like leaves.
Hyssopus (hiss-SO-pus). Hyssop.
Hystrix (HISS-triks). Bottle brush grass.

I

Ibericus (eye-BEER-ih-kus). Of Iberia (Spain, Portugal).
Iberis (eye-BEE-riss). Candytuft.
Ibota (eye-BOH-tuh). Japanese name for Ibota privet, sometimes specific name.
Idesia (eye-DEE-zee-uh). Tree, East Asia.
Idria (ID-ree-uh). Tree of Southern California.
Igneus (IG-nee-us). Fiery, flame-colored.
Ilex (EYE-leks). Holly.
Illicilfolius (ih-liss-ih-FO-lee-us). *Ilex*-leaved, holly-leaved.
Illicium (ih-LISS-ee-um). Star or Chinese anise.
Imbricatus (im-brih-KAY-tus). Overlappling, as shingles.
Immaculatus (ih-mak-yoo-LAY-tus). Immaculate, spotless.
Impatiens (im-PAY-she-enz). Balsam.
Imperata (im-pay-RAH-tuh). Imperial, regal.
Imperator (im-pur-AY-tor). Showy.
Imperialis (im-peer-ee-AY-liss). Imperial, kingly.
Incanus (in-KAY-nus). Hoary, grayish-white.
Incarnatus (in-kahr-NAY-tus). Flesh-colored.

Incarvillea (in-kahr-VIL-ee-uh). Hardy gloxinia.
Incisus (in-SY-sus). Cut.
Incomparabilis (in-kom-puh-RAB-ih-liss). Incomparable, excelling.
Indehiscent (in-dee-HISS-ent). Not splitting.
Indentatus (in-den-TAY-tus). Indented, toothed.
Indicus (IN-dih-kus). Of India.
Indigofera (in-dih-GOF-ur-uh). Indigo.
Indivisus (in-dih-VY-zus). Undivided.
Indocalamus (in-doh-KAL-ih-mus). Reed of India.
Inermis (in-UR-miss). Unarmed, without thorns.
Inflorescence (in-flo-RESS-enz). Arrangement of flowers on stem.
Inga (ING-guh). Tropical acacia-like trees.
Insignis (in-SIG-niss). Remarkable, distinguished.
Insularis (in-syoo-LAYR-us). Of an island.
Integrifolius (in-teg-rih-FO-lee-us; in-tee-grih-FO-lee-us). Leaves entire.
Intortus (in-TOR-tus). Twisted.
Intumescens (in-tyoo-MESS-enz). Swollen, puffed up, tumid.
Inula (IN-yoo-luh). Rather coarse hardy plants.
Involucre (IN-vo-lyoo-kur). Circle or collection of bracts.
Involute (IN-vo-lyoot). Rolled inward.
Iochroma (eye-o-KROW-muh). Violet-colored.
Ioensis (eye-o-EN-siss). Of Iowa, specific name of *Malus*.
Ionanthus (eye-o-NAN-thus). Violet-flowered.
Ionopsidium (eye-o-nop-SID-ee-um). Diamond flower.
Ipomoea (ip-o-MEE-uh; eye-po-MEE-uh). Cultivated morning-glory.
Iresine (ir-eye-SY-nee; eye-ree-SY-nee). Bid-leaf; bedding plants.
Iridescens (ir-ih-DESS-enz). Iridescent.
Iridifolius (eye-rih-dih-FO-lee-us). Iris-like foliage.
Iris (EYE-riss). Common garden perennials.
Isatis (EYE-suh-iss). Woad.

Ismene (is-MEE-nee). Peruvian daffodil, spider lily; now *Hymenocallis.*
Isoloma (eye-so-LO-muh). Showy greenhouse plants.
Isophyllus (eye-so-FIL-us). Leaves all the same form.
Isoplexis (eye-so-PLEK-siss). Subtropical undershrubs.
Isopogon (eye-so-PO-gon). Australian shrubs.
Italicus (ih-TAL-ih-kus). Of Italy.
Itea (IT-ee-uh, eye-TEE-uh). Sweet spire, Virginia willow.
Ixia (IK-see-uh). Cormous plants like gladioli.
Ixiolirion (iks-ee-o-LIR-ee-on). Hardy bulbous herbs.
Ixora (ik-SO-ruh). Jungle geranium, iron tree.

J

Jacaranda (jak-uh-RAN-duh). Showy tropical trees.
Jackmani (jak-MAN-eye). After proper name, species of *Clematis.*
Jacobinia (jak-o-BIN-ee-uh). Showy greenhouse plants.
Jamesia (JAYMZ-ee-uh). Rocky mountain shrubs.
Japonica (juh-PON-ih-kuh). Of Japan.
Jasione (jas-ee-O-nee). Rock garden herbs; sheep's bit.
Jasmine (JAZ-min; JASS-min). Common name for several plants.
Jasminoides (jaz-min-OY-deez). Resembling jasmine.
Jasminum (JAZ-mih-num). Jasmine or jessamine.
Jatropha (JAT-troh-fuh). Barbados nut, coral plant.
Javanicus (juh-VAN-ih-kus). Of Java.
Jeffersonia (jef-ur-SO-nee-uh). Twinleaf.
Johnsoni (jon-SO-nee). From proper name.
Jonquil (JON-kwil). One of small cluster narcissi.
Juglans (JOO-glanz). Walnut, butternut.
Juncus (JUNG-kus). True rush.
Juniper (JOO-nih-pur). Coniferous trees, shrubs.
Juniperus (joo-NIP-ur-us). Generic name of juniper.
Justicia (jus-TISH-ee-uh). Fairly showy greenhouse plants.

K

Kaempferi (KEM-fur-eye). From proper name.
Kalanchoe (kal-an-KO-ee). Showy succulents.
Kalmia (KAL-mee-uh). Mountain laurel, sheep laurel.
Kamtschaticus (kam-CHAT-ih-kus). Of Kamchatka, Siberia.
Kentia (KEN-tee-uh). Florists' palms.
Kerria (KER-ee-uh). Deciduous shrubs.
Keteleeria (kee-teh-LEE-ree-uh). Chinese evergreen trees.
Kewensis (kyoo-EN-siss). Referring to Kew Gardens.
Kinnikinnick (kin-ih-kih-NIK). Red bearberry or silky cornel.
Kniphofia (nip-HO-fee-uh; ny-FO-fee-uh). Torch lily, poker plant.
Kochia (KO-kee-uh). Summer cypress.
Koeleria (ko-LAYR-ee-uh). From proper name.
Koelreuteria (kel-roo-TEE-ree-uh). Goldenrain tree.
Kolkwitzia (kol-KWIT-zee-uh). Beauty-bush.
Koreanus (kor-ee-AY-nus). Of Korea, also coreanus.
Krameri (KRAY-mur-eye). From proper name.
Krigia (KRIG-ee-uh). Plants like dandelions.
Kuhnia (KOO-nee-uh). Small perennial herbs; from proper name.

L

Labiata (lay-bee-AY-tuh). Lipped.
Labiatae (lay-bee-AY-tee). Mint family.
Lablab (LOB-lob). Hyacinth bean; from Hindu name.
Laburnum (luh-BUR-num). Golden chain, bean tree.
Lachenalia (lak-eh-NAY-lee-uh). Cape cowslip, leopard lily.
Laciniatus (luh-sin-ee-AY-tus). Torn, fringed.
Lactatus (lak-TAY-tus). Milky.
Lacteus (LAK-tee-us). Milk-white.
Lactiflorus (lak-tih-FLO-rus). Milk-white flowers.
Lactuca (lak-TYOO-kuh). Lettuce.
Laelia (LEE-lee-uh). Showy florists' orchids.

Laeliocattleya (lee-lee-o-KAT-lee-uh). Bigeneric hybrid orchids.
Laevigatus (lev-ih-GAY-tus; lee-vih-GAY-tus). Smooth.
Laevis (LEE-viss). Smooth.
Lagenaria (laj-eh-NAY-ree-uh). Bottle, dipper, calabash gourd.
Lagerstroemia (lay-gur-STREE-mee-uh; luh-gur-STREE-mee-uh). Crape myrtle.
Lagurus (luh-GYOO-rus). Hare's-tail or rabbit's-tail grass.
Lamarckia (luh-MAHR-kee-uh). Goldentop.
Lambertia (lam-BURT-ee-uh). Australian shrubs.
Lamium (LAY-mee-um). Dead nettle.
Lampranthus (lam-PRAN-thus). With brilliant flowers.
Lanatus (luh-NAY-tus). Woolly.
Lanceolata (lan-see-o-LAY-tuh). Shaped like a lance head.
Lancifolium (lan-see-FO-lee-um). Leaves lance-shaped.
Lantana (lan-TAY-nuh). Subtropical shrubs, bedding plants.
Lantanoides (lan-tay-NOY-deez). Lantana-like.
Lanuginosus (luh-nyoo-jih-NO-sus; lan-yoo-jih-NO-sus). Woolly, downy.
Lapageria (lap-uh-JEE-ree-uh). Chilean bellflower, Chile bells.
Laurifolius (law-rih-FO-lee-us). Laurel-leaved.
Larix (LAR-iks; LAY-riks). Larch.
Larrea (LAR-ee-uh). Creosote bush, greasewood.
Lasiocarpus (lay-see-o-KAHR-pus). Rough or woolly-fruited.
Lasthenia (lass-THEE-nee-uh). West coast fleshy annuals.
Latania (la-TAY-nee-uh). Fan palms.
Lathyrus (LATH-ih-rus). Sweet pea.
Latiflorus (lat-ih-FLO-rus). Broad-flowered.
Latifolius (lat-ih-FO-lee-us). Broad-leaved.
Latisquamus (lat-ih-SKWAW-mus; lat-ih-SKWAY-mus). With broad scales or bracts.
Latus (LAY-tus). Broad, wide.
Laurus (LAW-rus). True laurel, bay tree, sweet bay.
Laurocerasus (law-ro-SEH-ra-sus). Cherry laurel, evergreen shrub of rose family.
Lavandula (luh-VAN-dyoo-luh). Lavender.

Lavatera (lav-uh-TEE-ruh). Tree mallow.
Lawsonia (law-SO-nee-uh). Mignonette tree, henna plant.
Lawsonianum (law-so-nee-AY-num). From proper name.
Laxiflorus (laks-ih-FLOR-us). Loose-flowered.
Laxifolius (laks-ih-FO-lee-us). Loose-leaved.
Laxus (LAKS-us). Loose.
Layia (LAY-ee-uh; LAY-ee-yuh). Tidytips, white daisies.
Ledebouria (leh-dih-BOR-ee-uh). Bulbous, perennial herbs; from proper name.
Ledum (LEE-dum). Labrador tea, wild rosemary.
Legume (LEG-yoom; le-GYOOM). Plant or fruit of pea family.
Leguminosae (leh-gyoo-mih-NO-see). Pea, bean or pulse family.
Leguminous (leh-GYOO-mih-nus). Pertaining to legumes.
Leichtlini (LIKED-lin-eye). From proper name.
Leiophyllum (ly-o-FIL-um). Sand myrtle.
Lemaireocereus (leh-may-ree-o-SEE-ree-us). Ribbed tree-like cacti.
Lemna (LEM-nuh). Duckweed.
Lemoinei (leh-MOY-nee-eye). From proper name.
Lentago (len-TAY-go). Early name for viburnum, species name of *Viburnum*.
Leonotis (lee-on-O-tiss). Resembling a lion's ear (corolla).
Leontopodium (lee-on-to-PO-dee-um). Edelweiss.
Lepachys (LEP-uh-kiss). Prairie coneflower.
Lepidium (leh-PID-ee-um). Peppergrass, pepper cress.
Lepidophyllus (lep-ih-doh-FIL-us). Scaly-leaved.
Lepidus (LEP-ih-dus). Graceful, elegant.
Leptocaulis (lep-toh-KAW-liss). Thin-stemmed.
Leptodermis (lep-toh-DUR-miss). Thin-skinned.
Leptolepis (lep-TOL-eh-piss). Thin-scaled.
Leptophyllus (lep-toh-FIL-us). Thin-leaved.
Leptospermum (lep-to-SPUR-mum). Australian tea tree.
Leptosyne (lep-TOS-ih-nee). Tall coreopsis-like plants.
Lespedeza (less-peh-DEE-zuh). Bush clover.

Leucadendron (lyoo-kuh-DEN-dron). Silver tree.
Leucaena (loo-KY-nuh). White-colored.
Leucanthemum (lyoo-KAN-thee-mum). Early name of white daisy.
Leucanthus (lyoo-KAN-thus). White-flowered.
Leucocarpus (loo-ko-KAHR-pus). White-fruited.
Leucocoryne (lyoo-ko-ko-REE-nee). Glory of the sun.
Leucocrinum (lyoo-KOK-rih-num; lyoo-ko-KRIH-num). Sand or star lily.
Leucodermis (loo-ko-DUR-miss). White-skinned.
Leucojum (lyoo-KO-jum). Snowflake.
Leucophyllum (lyoo-ko-FIL-um). Texas, Mexican shrubs.
Leucospermum (loo-ko-SPUR-mum). White-seeded.
Leucothoe (lyoo-KOTH-o-ee). Shrubs of heath family.
Levisticum (leh-VISS-tih-kum). Lovage.
Lewisia (lyoo-ISS-ee-uh). Fleshy rockery herbs; bitterroot.
Liatris (ly-AY-triss). Gayfeather, blazing star.
Libericus (ly-BEER-ih-kus). Of Liberia.
Libocedrus (ly-boh-SEE-drus). Incense cedar.
Lichen (LY-ken). Curious moss-like plants.
Licuala (lik-yoo-AY-luh). Small fan palms.
Ligularia (lig-yoo-LAY-ree-uh). Leopard plant.
Ligulatus (lig-yoo-LAY-tus). Strap-shaped.
Ligustrina (ly-gus-TRY-nuh). Flexible, privet-leaved.
Ligustrum (lih-GUS-trum). Privet; hedge plants.
Lilacinus (ly-LASS-ih-nus; ly-luh-SY-nus). Lilac color or form.
Lilium (LIL-ee-um). Lily.
Limnanthes (lim-NAN-theez). Meadow foam, marshflower.
Limnocharis (lim-NOK-uh-riss). Tropical aquatic herbs.
Limonia (ly-MO-nee-uh). Old generic name for lemon and lime.
Limonium (ly-MO-nee-um). Sea lavender.
Linanthus (ly-NAN-thus). Ground pink.
Linaria (ly-NAY-ree-uh). Toadflax.
Lindera (lin-DER-uh). Aromatic shrubs, trees of laurel family.
Linear (LIN-ee-ur). Long, narrow; parallel margins.

Lineatus (lin-ee-AY-tus). With lines or stripes.
Linifolius (lin-ih-FO-lee-us). Flax-leaved.
Lingulatus (ling-yoo-LAY-tus). Tongue-shaped.
Linnaea (lih-NEE-uh). Twinflower; trailing evergreens of honeysuckle family.
Linosyris (ly-NOSS-ih-riss). Goldilocks.
Linus (LY-nus). Flax.
Liparis (LIP-uh-riss). Twayblade, orchids.
Lippia (LIP-ee-uh). Lemon verbena.
Liquidambar (lik-wid-AM-bahr). Sweet gum.
Liriodendron (lih-ree-o-DEN-dron). Tulip tree.
Liriope (lih-RY-o-pee). Lily-turf.
Listera (LISS-tur-uh; liss-TEE-ruh). Hardy terrestrial orchids.
Litchi (LEE-chee). Chinese fruit tree.
Lithocarpus (lith-o-KAHR-pus). Tanbark oak.
Lithops (LITH-ops). Stoneface, stone plant.
Lithospermum (lith-o-SPUR-mum). Gromwell, puccoon.
Litsea (LIT-see-uh). Aromatic trees, shrubs.
Littoralis (lih-tor-AY-liss). Of the seashore.
Livistona (liv-ih-STOH-nuh). Chinese, Australian fan palms.
Lobatus (lo-BAY-tus). Lobed.
Lobe (LOHB). Any segment, especially if rounded.
Lobelia (lo-BEE-lee-uh; lo-BEEL-yuh). Popular garden plants.
Lobularia (lo-byoo-LAY-ree-uh). Sweet alyssum.
Loiseleuria (loy-seh-LYOO-ree-uh). Alpine azaleas.
Lolium (LO-lee-um; LOL-ee-um). Rye grass.
Lomaria (lo-MAY-ree-uh). Ferns.
Lomatia (lo-MAY-tee-uh; lo-MAY-she-uh). Crinkle bush.
Longiflorum (lon-jih-FLO-rum). With long flowers.
Longifolius (long-ih-FO-lee-us). Long-leaved.
Longipes (LON-jih-peez). Long-footed or stalked.
Longissimus (lon-JISS-ih-mus). Longest.
Lonicera (lo-NISS-ur-uh). Honeysuckle.
Lopezia (lo-PEE-zee-uh). Shrubby greenhouse plants.
Loquat (LO-kwat; LO-kwot). Asiatic evergreen trees.

Lotus (LO-tus). Shrubby and forage plants.
Lowei (LO-ee-eye). From proper name.
Lucidus (LYOO-sih-dus). Bright, shining, clear.
Luculia (lyoo-KYOO-lee-uh). Showy greenhouse shrubs.
Lucuma (lyoo-KYOO-muh). Sapote, eggfruit, canistel.
Ludwigia (lud-WIG-ee-uh). False or swamp loosestrife.
Luffa (LUF-uh). Dishcloth gourd.
Luma (LOO-muh). From Chilean name.
Lunaria (lyoo-NAY-ree-uh). Honesty, moonwort.
Lupinus (lyoo-PY-nus). Lupine.
Luridus (LYOO-rih-dus). Pale yellow; firelike.
Lutescens (lyoo-TESS-enz). Becoming yellowish.
Luteus (LYOO-tee-us). Yellow.
Luzula (loo-ZOO-luh; loo-ZYOO-luh). From Italian vernacular.
Lycaste (ly-KASS-tee). Showy greenhouse orchids.
Lychnis (LIK-niss). Catchfly, campions.
Lycium (LISS-ee-um). Matrimony vine, boxthorn.
Lycopersicon (ly-ko-PUR-sih-kon). Tomato.
Lycopodium (ly-ko-PO-dee-um). Club moss.
Lycopus (LY-ko-pus). Water horehound.
Lycoris (ly-KO-riss). Hardy amaryllis.
Lygodium (ly-GO-dee-um). Climbing fern.
Lyonia (ly-O-nee-uh). Shrubs of heath family.
Lyrata (ly-RAY-tuh). Lyre-shaped.
Lysimachia (ly-sih-MAY-kee-uh; liss-ih-MAK-ee-uh). Loosestrife.
Lythrum (LITH-rum). Purple loosestrife.

M

Maackia (ma-AY-kee-uh). Asiatic trees, shrubs.
Macafeeana (mak-af-ee-AY-nuh). From proper name.
Macfadyena (mak-fad-ee-EN-uh). From proper name.
Machaeranthera (mak-ur-an-THAYR-uh). With sword-like flowers.
Mackaya (ma-KAY-uh). Tender shrubs.
Macleaya (mak-LAY-uh). Plume poppy.

Maclura (muh-KLYOO-ruh). Osage orange.
Macracanthus (mak-ruh-KAN-thus). Large-spined.
Macranthus (muh-KRAN-thus). Large-flowered.
Macrocarpus (mak-ro-KAHR-pus). Large-fruited.
Macrocephalus (mak-ro-SEF-a-lus). Large-headed.
Macrophyllus (mak-ro-FIL-us). Large-leaved.
Macrostachya (mak-ro-STAK-ee-uh). Great-spiked.
Maculatus (mak-yoo-LAY-tus). Spotted.
Maculosa (mak-yoo-LO-suh). Spotted.
Madagascariensis (ma-dah-GAS-kar-ee-EN-siss). Of Madagascar.
Madia (MAY-dee-uh). Tarweed.
Magellanicus (ma-jel-AN-ih-kus). Of Straits of Magellan.
Magnificus (mag-NIF-ih-kus). Magnificent, distinguished.
Magnolia (mag-NO-lee-uh). Showy trees, shrubs.
Mahaleb (MAH-huh-leb; muh-HA-leb). A European cherry.
Mahernia (may-HUR-nee-uh). Honey bells.
Mahoberberis (ma-ho-BEHR-behr-iss). Bigeneric name for half-evergreen shrub, hybrid between *Mahonia aquifolium* and *Berberis vulgaris*.
Mahonia (ma-HO-nee-uh). Oregon grape.
Maianthemum (may-YAN-thee-mum). False lily of the valley.
Majalis (muh-JAY-liss). Of May, Maytime.
Major (MAY-jor). Greater, larger.
Majorana (may-jo-RAY-nuh). Sweet marjoram.
Malacoides (mal-uh-KOY-deez). Soft, mucilaginous.
Malcomia (mal-KO-mee-uh). Malcolm or Virginia stocks.
Malope (MAL-o-pee). Showy annuals.
Malpighia (mal-PIG-ee-uh). Barbados cherry.
Malus (MAY-lus). Apple.
Malva (MAL-vuh). Mallow.
Malvastrum (mal-VAS-trum). False or prairie mallow.
Malvaviscus (mal-vuh-VISS-kus). Turk's-cap.
Mammillaria (mam-ih-LAY-ree-uh). Cylindrical and spherical cacti.
Mandevilla (man-deh-VIL-uh). Tropical woody vines.

Mandragora (man-DRAG-or-uh). Mandrake.
Manetti (muh-NET-tee). Rose understock.
Manettia (muh-NET-ee-uh). Showy greenhouse vines.
Manfreda (man-FREE-duh). False aloe.
Mangifera (man-JIFF-ur-uh). Mango.
Manicata (man-ih-KAY-tuh). Long-sleeved.
Manihot (MAN-ih-hot). Cassava, tapioca plant.
Maranta (muh-RAN-tuh). Foliage plants; arrowroot.
Marchantia (mahr-KAN-she-uh; mahr-KAN-tee-uh). Liverwort.
Marginalis (mahr-jih-NAY-liss). Margined, usually different color.
Marginatus (mahr-jih-NAY-tus). Margined or striped.
Marguerite (mahr-geh-REET). A white daisy.
Mariana (mayr-ee-AY-nuh). Maryland.
Marica (MAYR-ih-kuh). Tropical iris-like plants.
Marilandicus (mayr-ih-LAN-dih-kus). Of the Maryland region; also written marylandicus.
Maritimus (muh-RIT-ih-mus). Of the sea or shore.
Marmorata (mahr-mo-RAY-tuh). Mottled.
Marrubium (mah-ROO-bee-um). Horehound.
Marsilea (mahr-SIL-ee-um). Aquatic herbs; pepperwort.
Martagon (MAHR-tuh-gon). A kind of turban.
Martinezia (mahr-tih-NEE-zee-uh). Spiny feather palms.
Martynia (mahr-TIN-ee-uh). Unicorn plant, proboscis flower.
Mas (MAS). Male.
Masculatus (mas-kyoo-LAY-tus). Masculine.
Masdevallia (mas-deh-VAL-ee-uh). Tropical American orchids.
Massangeana (mah-sahn-jee-AY-nuh). From proper name.
Mathiola (ma-thee-O-luh). Stock, gillyflower.
Matricaria (mat-rih-KAY-ree-uh). Chamomile.
Matronalis (ma-tro-NAY-liss). Sedate, often hoary.
Matthiola (ma-tee-O-luh). Annual and perennial herbs. From proper name.
Maurandia (maw-RAN-dee-uh). Tender climbers, trailers.
Mauritia (maw-RISH-ee-uh). Ita palm.
Maxillaria (mak-sih-LAY-ree-uh). Tropical American orchids.

Maximiliana (mak-sih-mil-ee-AY-nuh). Tall feather palms.
Maximus (MAKS-ih-mus). Largest.
Maytenus (MAY-teh-nus). From Chilean vernacular.
Mazus (MAY-zus). Flowering ground cover plants.
Meconopsis (mee-ko-NOP-siss). Poppy-like plants.
Medeola (mee-DEE-o-luh). Indian cucumber.
Media (MEE-dee-uh). Intermediate.
Medicago (med-ih-KAY-go). Alfalfa.
Medinilla (med-ih-NIL-uh). Striking old greenhouse plants.
Medullaris (med-yoo-LAY-riss). Of the marrow or pith.
Melaleuca (mel-uh-LYOO-kuh). Bottle brush.
Melanthium (mee-LAN-thee-um). Bunchflower.
Melastoma (mee-LASS-toh-muh). Asiatic shrubs.
Meleagris (mel-ee-AY-griss). Speckled.
Melia (MEE-lee-uh). China tree, Texas umbrella tree.
Melianthus (mel-ee-AN-thus). Honeyflower, honey bush.
Melica (MEE-lee-kuh). Perennial grass.
Melilotus (mel-ee-LO-tus). Sweet clover.
Melissa (meh-LISS-uh). Attracts honeybees.
Melocactus (mel-o-KAK-tus). Apple-shaped cactus.
Mendelli (men-DEL-eye). From proper name.
Menispermum (men-ih-SPUR-mum). Moonseed.
Mentha (MEN-thuh). Mint.
Mentzelia (ment-ZEE-lee-uh). Prairie lily, blazing star.
Menyanthes (men-ee-AN-theez). Bogbean, buckbean.
Meratia (mee-RAY-she-uh). Winter-flowering shrubs.
Mertensia (mur-TEN-see-uh). Bluebells.
Mesembryanthemum (mess-em-bree-AN-thee-mum; mee-zem-bree-AN-thee-mum). Fig marigold.
Mespilus (MESS-pih-lus). Medlar.
Mesquite (mess-KEET; MESS-keet). Desert shrub.
Metallica (meh-TAL-ih-kuh). With metallic sheen.
Metasequoia (met-uh-sih-KWOY-uh). Generic name of Dawn redwood.

Metrosideros (mee-tro-sih-DEE-ross; met-ro-sih-DEE-ross). Iron tree.
Mexicana (mek-sih-KAY-nuh). Of Mexico.
Mezereum (meh-ZEE-ree-um). Persian name.
Micans (MY-kanz). Glittering, sparkling.
Michauxia (mih-SHAWK-see-uh; mih-SHOW-ee-uh). Showy border plants.
Michelia (my-KEE-lee-uh). Banana shrub.
Miconia (my-KO-nee-uh). Showy foliage plants.
Micranthus (my-KRAN-thus). Small-flowered.
Microcarpus (my-kro-KAHR-pus). Small-fruited.
Microcitrus (my-kro-SIT-rus). Finger lime.
Micromeria (my-kro-MEE-ree-uh). Yerba buena.
Microphylla (my-kro-FIL-uh). Small-leaved.
Midrib (MID-rib). Main vein or rib of leaf.
Mignonette (min-yun-ET). Fragrant garden annuals.
Milla (MIL-uh). Mexican star, frost flower.
Millettia (mil-LET-ee-uh). From proper name.
Miltonia (mil-TOH-nee-uh). Pansy orchids.
Mimosa (mih-MO-suh). Sensitive plants.
Mimulus (MIM-yoo-lus). Monkey flower.
Miniatus (min-ee-AY-tus). Cinnabar-red.
Minima (MIN-ih-muh). Smallest, dwarf.
Minor (MY-nor). Smaller.
Mirabilis (mih-RAB-ih-liss; my-RAB-ih-liss). Four-o'clocks; also extraordinary.
Miscanthus (miss-KAN-thus). Eulalia.
Mistletoe (MISS-il-toh). Tree parasite.
Mitchella (mih-CHEL-uh). Partridgeberry, twinberry.
Mitella (mih-TEL-uh). Bishop's-cap, fairy cap.
Molinia (mo-LIN-ee-uh). Perennial grasses.
Mollis (MOL-liss). Soft, hairy.
Mollissimus (mol-LISS-ih-mus). Very soft, hairy.
Molucella (mo-lyoo-SEL-uh). Shellflower, molucca balm.
Momordica (mo-MOR-dih-kuh). Balsam apple and pear.

Monadelphus (mon-uh-DEL-fus). In one bundle.
Monarda (mo-NAR-duh). Bee balm, horsemint, bergamot.
Monocotyledonous (mon-o-kot-ih-LEE-dun-us). Having but one cotyledon.
Monoecious (mo-NEE-shus). Stamens and pistils in separate flowers on same plant.
Monotropa (mo-NOT-ro-puh). Indian pipe.
Monstera (MON-stur-uh; mon-STEE-ruh). Huge tropical climbers.
Monstrosus (mon-STRO-sus). Monstrous, abnormal.
Montanus (mon-TAY-nus). Pertaining to mountains.
Montia (MON-tee-uh). Winter purslane, Indian lettuce.
Monticolus (mon-TIK-o-lus). Inhabiting mountains.
Morea (mo-REE-uh). Tender iris-like plants.
Morello (mo-REL-o). A kind of sour cherry.
Morifolium (mor-ih-FO-lee-um). With mulberry-like leaves.
Morinda (mor-IN-duh). Indian mulberry.
Morus (MO-rus). Mulberry.
Mosaicus (mo-ZAY-ih-cus). Variegated.
Moschata (moss-KAY-tuh). Musky, musk-scented.
Moutan (MOO-tan). Chinese name for tree peony.
Mucosus (myoo-KO-sus). Sticky, slimy.
Mucronata (myoo-krow-NA-tuh). With small, abrupt tip.
Muehlenbeckia (myoo-len-BEK-ee-uh). Hanging basket plants.
Mughus (MYOO-gus). Form of mugo; mountain pine.
Multicaulis (mul-tih-KAW-liss). Many-stemmed.
Multicolor (MUL-tih-kul-ur). Many-colored.
Multifidus (mul-TIF-ih-dus). Many times parted or cleft.
Multiflorus (mul-tih-FLO-rus). Many-flowered.
Multijuga (mul-tih-JYOO-guh). In many pairs.
Muralis (myoo-RAY-liss). Growing on walls.
Muricatus (myur-ih-KAY-tus). Roughened by sharp points.
Murraya (mur-RAY-uh). From proper name.
Musa (MYOO-zuh). Banana.
Muscari (mus-KAY-rye). Grape hyacinth.

Muscosus (myoos-KO-sus). Moss-like.
Mutabilis (myoo-TAB-ih-liss). Variable.
Myoporum (my-OP-or-um). Evergreen trees and shrubs with spotted leaves.
Myosotis (my-o-SO-tiss). Forget-me-not.
Myrica (mih-RY-kuh). Bayberry, wax myrtle.
Myrica (MEER-ih-kuh). Greek name for Tamarisk.
Myricaria (mir-ih-KAY-ree-uh). False tamarisk.
Myriocephalus (mir-ee-o-SEF-uh-lus). An annual everlasting.
Myriophyllum (mir-ee-o-FIL-um). Parrot's-feather, water millfoil.
Myrrhis (MIR-iss). Myrrh, sweet cicely.
Myrtifolius (mur-tih-FO-lee-us). Myrtle-leaved.
Myrtilloides (mir-til-LOY-deez). Myrtle-like.
Myrtus (MUR-tus). True myrtle.

N

Naegalia (nay-JEE-lee-uh). Flowering potted plants.
Nana (NAN-uh). Dwarf.
Nandina (nan-DY-nuh). Red-fruited, fine-textured evergreen shrub.
Nanus (NAY-nus). Dwarf.
Napellus (nuh-PEL-us). Roots like turnip.
Narcissus (nahr-SISS-us). Daffodil, jonquil.
Nasturtium (nuh-STUR-shum). Common garden annuals.
Natans (NAY-tanz). Floating, swimming.
Neapolitanum (nee-a-pol-ih-TAY-num). Of Naples.
Nectariferous (nek-tur-IF-ur-us). Producing nectar.
Neglectus (neg-LEK-tus; neh-GLEK-tus). Neglected, overlooked.
Negundo (nee-GUN-doh). Species name of *Acer*.
Neillia (NEEL-ee-uh). Shrubs like spireas.
Nelumbium (nee-LUM-bee-um). Lotus.
Nemastylis (ne-MASS-tih-liss). Tender bulbous plants.
Nemesia (neh-MEE-she-uh; neh-MEE-see-uh). Garden plants.
Nemopanthus (nee-mo-PAN-thus). Mountain holly.

Nemophila (nee-MOF-ih-luh). Baby blue-eyes.
Nemoralis (nem-o-RAY-liss). Of groves or woods.
Nepenthes (nee-PEN-theez). Pitcher plant.
Nepeta (NEP-eh-tuh; neh-PEE-tuh as common name). Rockery plants.
Nephrolepis (nee-FROL-eh-piss; neh-FROL-eh-piss). Boston or sword fern.
Nerine (nee-RY-nee). Guernsey lily.
Nerium (NEE-ree-um). Oleander.
Nertera (nur-TEE-ruh). Bead plant.
Nervosus (nur-VO-sus). Veined or nerved.
Neviusia (nev-ih-YOO-see-uh). Snow wreath; from proper name.
Nicandra (ny-KAN-druh). Apple-of-Peru.
Nicotiana (nih-ko-she-AY-nuh). Tobacco, flowering tobacco.
Nidularium (nid-yoo-LAY-ree-um). Nested (flowers).
Nidus (NY-dus). Nest.
Nierembergia (nee-rem-BUR-gee-uh; nee-rem-BUR-jee-uh). Cupflower.
Nigella (ny-JEL-uh). Love-in-a-mist.
Niger (NY-jur). Black.
Nigra (NY-gruh). Black.
Nigrescens (ny-GRESS-enz). Blackening.
Nigricans (NY-grih-kanz). Black.
Nitidus (NIT-ih-dus). Shining.
Nivalis (nih-VAY-liss). Snowy, white.
Niveus (NIH-vee-us). White, snowy.
Nobilis (NO-bil-us). Noble, famous.
Noctiflora (nok-tih-FLO-ruh). Night-blooming.
Nocturnus (nok-TUR-nus). Of the night.
Nodose (NO-dohs; no-DOHS). Knotty or knobbed.
Nodule (NOD-yool). Lump, protuberance.
Noisette (nwuh-ZET). Race of hardy roses.
Nolana (no-LAY-nuh). Chilean bellflower.
Nolina (no-LY-nuh). Bear grass.
Nomenclature (NO-men-KLAY-tyur). System of naming.

Nopalea (no-PAY-lee-uh). Opuntia-like cacti.
Notholcus (no-THOL-kus). Velvet grass.
Nothofagus (no-tho-FAY-gus). False beech.
Novae-angliae (no-vee-AN-glih-ee; no-veh-AN-glih-ee). Of New England.
Novi-belgi (no-vih-BEL-jy). Of New York.
Nuciferus (nyoo-SIF-ur-us). Nut-bearing.
Nudatus (nyoo-DAY-tus). Nude, stripped.
Nudiflora (nyoo-dih-FLO-ruh). Flowers without leaves.
Nudum (NYOO-dum). Bare, naked.
Nutans (NYOO-tanz). Nodding.
Nyctanthes (nik-TAN-theez). Night jasmine.
Nymphaea (nim-FEE-uh). Water lily.
Nymphoides (nim-FOY-deez). Floating heart.
Nymphozanthus (nim-fo-ZAN-thus). Coarse water plants.
Nyssa (NISS-uh). Tupelo, black gum, sour gum.

O

Obconica (ob-KON-ih-kuh). Inversely conical.
Obcordate (ob-KOR-dayt). Inverted heart-shaped.
Obliquus (o-BLY-kwus, o-BLIH-kwus). Lopsided, diagonal.
Oblongifolia (ob-long-ih-FO-lee-uh). Oblong-leaved.
Oblongus (ob-LONG-us). Oblong.
Obovata (ob-o-VAY-tuh). Inversely ovate.
Obscurus (ob-SKYUR-us). Obscure, hidden.
Obtuse (ob-TYOOS). Blunt or rounded at end.
Obtusifolius (ob-too-sih-FO-lee-us). Blunt-leaved.
Occidentalis (ok-sih-den-TAY-liss). Western, New World.
Ocellata (oss-eh-LAY-tuh). With small eyes.
Ochna (OAK-nuh). Bird's eye bush.
Ochraceus (o-KRAY-shus). Pale yellow.
Ochroleucus (ok-ro-LYOO-kus). Yellowish white.
Ochrosia (o-KRO-see-uh). Pale yellow.
Ocimum (O-see-mum). Fragrant.

Ocotillo (o-ko-TEE-yo; o-ko-TEEL-yo). Candlewood; *fouquieria*.
Oculatus (ok-yoo-LAY-tus). Eyed.
Odontioda (o-don-tee-O-duh). Bigeneric orchids.
Odontoglossum (o-don-toh-GLOSS-um). Showy spray orchids.
Odoratissimus (o-dor-uh-TISS-ih-mus). Very fragrant.
Odoratus (o-dor-AY-tus). Fragrant.
Oenothera (ee-no-THEE-ruh; ee-NOTH-ur-uh). Evening primrose.
Officinalis (o-fiss-ih-NAY-liss). Medicinal.
Olea (O-lee-uh). Olive.
Oleaceae (o-lee-AY-see-ee). Olive, lilac, ash family.
Oleander (o-lee-AN-dur). Nerium.
Olearia (o-lee-AY-ree-uh). Tree aster, daisy tree.
Oleifera (o-lee-IF-ur-uh; o-leh-IF-ur-uh). Oil-bearing.
Oleraceous (ol-ur-AY-shus). Edible.
Olericulture (OL-ur-ih-kul-tyur; o-LER-ih-kul-tyur as common name). Culture of vegetables.
Olivaceus (ol-ih-VAY-shus). Olive-like.
Olneya (OL-nee-uh). Desert ironwood.
Omorika (o-mor-EE-kuh). Serbian vernacular name; species of spruce.
Omphalodes (om-fuh-LO-deez). Navelwort.
Oncidium (on-SID-ee-um). Butterfly orchid.
Onoclea (on-o-KLEE-uh). Sensitive fern.
Ononis (o-NO-niss). Clover-like herbs; restharrow.
Onosma (o-NOSS-muh). Rockery plants.
Onychium (o-NIK-ee-um). Claw fern.
Ophioglossaum (off-ee-o-GLOSS-um; o-fee-o-GLOSS-um). Adder's-tongue fern.
Ophiopogon (o-fee-o-PO-gon; off-ee-o-PO-gon). Lily-turf, jaburan.
Opulifolius (op-yoo-lih-FO-lee-us). Opulus-leaved.
Opulus (OP-yoo-lus). Snowball, species of viburnum.
Opuntia (o-PUN-she-uh). Prickly pear; common cactus.
Orbiculatus (or-bik-yoo-LAY-tus). Orbicular, round.
Orchid (OR-kid). Any plant of the Orchidaceae.

Orchidaceae (or-kih-DAY-see-ee). Orchid family.
Orchis (OR-kiss). Woodland orchids.
Oregana (or-ee-GAY-nuh). Of Oregon.
Oreodoxa (o-ree-o-DOK-suh). Royal, cabbage palm.
Orientalis (or-ee-en-TAY-liss). Oriental, Eastern.
Origanum (o-RIG-uh-num). Pot or wild marjoram.
Ornatus (or-NAY-tus). Ornate, adorned.
Ornithogalum (or-nih-THOG-uh-lum). Star-of-Bethlehem.
Orontium (o-RON-she-um). Golden club, floating arum.
Oroxylon (o-ROKS-ih-lon). Indian trumpet flower.
Orthocarpus (or-thoh-KAHR-pus). Owl's-clover.
Oryza (o-RY-zuh). Rice.
Osmanthus (oss-MAN-thus; oz-MAN-thus). Holly olive, devilwood, tea olive; fragrant-flowered.
Osmunda (oss-MUN-duh; oz-MUN-duh). Royal, cinnamon fern.
Osteospermum (oss-tee-o-SPUR-mum). With hard seeds.
Ostrowskia (os-TRO-skee-uh). Giant bellflower; from proper name.
Ostrya (OSS-tree-uh). Hop hornbeam, ironwood.
Otaksa (o-TAK-suh). Japanese name for hydrangea.
Othonna (o-THON-uh). Hanging basket plants.
Ovary (O-vuh-ree). Part of pistil containing ovules.
Ovata (o-VAY-tuh). Egg-shaped; broad end at base.
Ovifera (o-VIF-ur-uh). Egg-bearing.
Ovule (O-vyool). Egg cell.
Oxalis (OK-suh-liss). Wood sorrel.
Oxyacanthus (ok-see-uh-KAN-thus). Sharp-spined.
Oxydendrum (ok-see-DEN-drum). Sourwood, sorrel tree.
Oxytropis (ok-SIT-ro-piss). Locoweed.

P

Pachira (pa-KY-ruh). From proper name.
Pachistima (puh-KISS-tih-muh). Low evergreen shrubs, or ground covers.

Pachycereus (pak-ih-SEE-ree-us). Tree-like cactus.
Pachyrhizus (pak-ih-RY-zus). Herbs with thick, tuberous roots.
Pachysandra (pak-ih-SAN-druh). Evergreen ground cover plants.
Pachystachys (puh-KISS-tuh-kiss). Jacobinia-like plants.
Pacificus (puh-SIF-ih-kus). Of the Pacific.
Paeonia (pee-O-nee-uh). Peony.
Paliurus (pal-ih-YOO-rus). Christ's-thorn.
Pallidus (PAL-lid-us). Pale.
Palmaceae (pal-MAY-see-ee). Palm family.
Palmata (pal-MAY-tuh). Leaf radiately lobed or divided.
Paludosus (pal-yoo-DOH-sus). Marsh-loving.
Palustris (pa-LUS-triss). Growing in the marsh.
Panax (PAY-naks). Ginseng.
Pancratium (pan-KRAY-she-um). Sea daffodil.
Pandanus (pan-DAY-nus). Screw pine.
Pandorea (pan-DOH-ree-uh). Bower plant.
Panduratus (pan-dyoo-RAY-tus). Fiddle-shaped.
Panicle (PAN-ih-kul). Loose, irregularly compound flower cluster.
Paniculatus (pan-ik-yoo-LAY-tus). Compound raceme.
Panicum (PAN-ih-kum). Millet, switch grass.
Pansy (PAN-zee). Certain violas.
Papaver (puh-PAY-vur). Poppy.
Papilio (puh-PIL-ee-o). A butterfly; species of *Oncidium*.
Papilionaceous (puh-pil-ee-o-NAY-shus). Flower, as in the pea.
Papillose (PAP-ih-lohss). Minute nipple-shaped projections.
Pappus (PAP-us). Tuft of hairs attached to seed.
Papyrifera (pap-ih-RIF-ur-uh). Paper-bearing, species of birch.
Papyrus (puh-PY-rus). A sedge used to make paper.
Paradisea (par-uh-DY-see-uh; par-uh-DISS-ee-uh). St.-Bruno's-lily.
Pardalinus (pahr-duh-LY-nus). Spotted.
Parkinsonia (pahr-kin-SO-nee-uh). Jerusalem thorn.
Parnassia (pahr-NASS-ee-uh). Grass-of-Parnassus.
Parochetus (puh-ROK-ee-tus). Blue oxalis.
Paronychia (par-o-NIK-ee-uh). Whitlowwort.

Paronychoides (puh-ron-ih-KOY-deez). Resembling whitlowwort.
Parrotia (par-RO-tee-uh). Shrubs like witch hazel.
Parthenium (pahr-THEE-nee-um). Desert undershrub; guayule.
Parthenocissus (pahr-theh-no-SISS-us). Woodbine, Boston ivy, Virginia creeper.
Parviflorus (pahr-vih-FLO-rus). Small-flowered.
Parvifolia (pahr-vih-FO-lee-uh). Small-leaved.
Parvum (PAHR-vum). Small.
Paspalum (PASS-puh-lum). Ornamental and forage grasses.
Passiflora (pass-ih-FLOR-uh). Passionflower.
Pastinaca (pass-tih-NAY-kuh). Parsnip.
Patens (PAY-tenz). Spreading.
Patulus (PAT-yoo-lus). Spreading.
Pauciflorus (paw-sih-FLO-rus). Few-flowered.
Paulinia (paw-LIN-ee-uh). Hothouse foliage plant.
Paulownia (paw-LO-nee-uh). Hardy Chinese trees.
Pavonia (puh-VO-nee-uh). Greenhouse potted plants.
Pectinata (pek-tih-NAY-tuh). Like a comb.
Pedata (peh-DAY-tuh). Pedate palmately divided, with segments cleft like bird's foot.
Pedicularis (peh-dik-yoo-LAY-riss). Wood betony, lousewort.
Pedilanthus (peh-dih-LAN-thus). Redbird cactus, slipperflower.
Peduncle (pee-DUNG-kul). Primary flower stalk.
Pedunculatus (peh-dunk-yoo-LAY-tus). Stalked.
Pelargonium (pel-ahr-GO-nee-um). Florists' geranium.
Pelecyphora (pel-eh-SIFF-o-ruh). Hatchet cactus.
Pellaea (peh-LEE-uh). Cliff brake.
Peltandra (pel-TAN-druh). Arrow arum.
Peltaria (pel-TAY-ree-uh). Shieldwort.
Peltata (pel-TAY-tuh). Like a shield.
Peltiphyllum (pel-tih-FIL-um). Umbrella plant.
Peltophorum (pel-TOFF-or-um). Shield-bearing (stigma).
Pelviformis (pel-vih-FOR-miss). Pelvis-shaped.
Pendulus (PEN-dyoo-lus). Hanging.

Peniocereus (pen-ee-o-SEE-ree-us). Deerhorn cactus.
Pennisetum (pen-ih-SEE-tum). Fountain grass, pearl millet.
Pennsylvanicum—or pensylvanicum (pen-sil-VAY-nih-kum). Of Pennsylvania.
Pentagyna (pen-tuh-JY-nuh). Five-fruited.
Pentandrus (pen-TAN-drus). Of five stamens.
Pentaphyllus (pen-tuh-FIL-us). Five-petaled.
Penstemon (pen-STEE-mon; PEN-stee-mon as common name). Beardtongue.
Peony (PEE-o-nee). Plant or flower of *Paeonia*.
Peperomia (pep-ur-O-mee-uh). Exotic, foreign.
Peregrinus (peh-reh-GRY-nus). Fine foliage house plants.
Perennis (pur-EN-iss). Perennial, lasting year after year.
Pereskia (pee-RESS-kee-uh). Shrubby, leafy cacti; from proper name.
Perfoliatus (pehr-fo-lee-AY-tus). Leaf surrounding the stem.
Perianth (PEH-ree-anth). Floral envelope, calyx and corolla.
Pericarp (PEH-rih-karp). Matured or ripened ovary wall.
Perilla (pee-RIL-uh). Coleus-like plant; beefsteak plant.
Periploca (peh-RIP-lo-kuh). Silk vine.
Peristerya (peh-rih-STEE-ree-uh). Dove orchid, Holy Ghost flower.
Pernettia (pur-NET-ee-uh). Low evergreen shrubs of heath family.
Pernettya (pur-NET-ee-uh). Ornamental shrubs, from proper name.
Perovskia (peh-ROF-skee-uh). Asian herbs, subshrubs; from proper name.
Perpusilla (pur-pyoo-SIL-luh). Very small.
Persea (PUR-see-uh). Avocado, red or bull bay.
Persica (PUR-sih-kuh). Of Persia; once genus of peach.
Persicifolius (pur-sik-ih-FO-lee-us). Peach-like leaves.
Persolutus (pur-so-LYOO-tus). Like a garland.
Pescatoria (pess-kuh-TOH-ree-uh). Tropical American orchids.
Petal (PET-uhl). A section of the corolla.

Petalostemum (pet-uh-lo-STEE-mum). Prairie clover.
Petiolaris (pet-ee-o-LAY-riss). With a leafstalk.
Petiole (PET-ee-ohl). Footstalk of leaf.
Petrea (PEE-tree-uh; peh-TREE-uh). Purple wreath.
Petrocoptis (peh-tro-KOP-tiss). Rockery plants.
Petrorhagia (pet-ro-RAH-gee-uh). Broken rock.
Petroselinum (peh-tro-seh-LY-num). Parsley.
Petunia (pee-TYOO-nee-uh). Popular garden annuals.
Phacelia (fuh-SEE-lee-uh). California bluebell.
Phaedranthus (fee-DRAN-thus). Handsome evergreen vine.
Phaeus (FEE-us). Dusky.
Phaius (FAY-ee-us). Large, showy orchids.
Phalaenopsis (fal-eh-NOP-siss). Moth orchid.
Phalaris (FAL-uh-riss). Canary and ribbon grass.
Phaseolus (fuh-SEE-o-lus). Bean.
Phellodendron (fel-o-DEN-dron). Cork tree.
Phellos (FEL-ohs). Cork.
Philadelphus (fil-uh-DEL-fus). Mock orange.
Philodendron (fil-o-DEN-dron). Popular indoor vine.
Phleum (FLEE-um). Timothy.
Phlomis (FLO-miss). Jerusalem sage.
Phlox (FLOKS). Showy garden plants.
Phoeniceus (feh-NISS-ee-us). Purple-red.
Phoenix (FEE-niks). Ornamental and date palms.
Phormium (FOR-mee-um). New Zealand flax.
Photinia (fo-TIN-ee-uh). Hardy Asian shrubs.
Phragmites (frag-MY-teez). Common reed grass.
Phrynium (FRY-nee-um). Hothouse foliage plants.
Phygelius (fy-JEE-lee-us). Cape fuchsia.
Phyllanthus (fih-LAN-thus). Gooseberry tree, myrobalan.
Phyllitis (fih-LY-tiss). Hart's-tongue fern.
Phyllostachys (fil-o-STAY-kiss; fil-o-STAK-iss). Bamboo-like grass.
Physalis (FY-suh-liss; FISS-uh-liss). Chinese lantern plant.
Physocarpus (fy-so-KAHR-pus). Ninebark.

Physostegia (fy-so-STEE-jee-uh). False dragonhead.
Phyteuma (fih-TYOO-muh). Horned rampion.
Phytolacca (fy-toh-LAK-uh). Pokeweed, pokeberry.
Picea (PY-see-uh; PIH-see-uh). Spruce.
Picta (PIK-tuh). Painted or variegated.
Picturatus (pik-tyoo-RAY-tus). Painted-leaved, variegated.
Pieris (PY-ur-iss). Broad-leaved evergreen shrubs.
Pilea (PY-lee-uh). Artillery plant.
Pileatus (py-lee-AY-tus). With a cap.
Pilifera (py-LIFF-ur-uh). Bearing soft hairs.
Pilose (PY-lohs). Hairy, especially soft hairs.
Pimelea (pih-MEE-lee-uh). Rice flower.
Pimenta (pih-MEN-tuh). Bay rum tree, allspice tree.
Pimpinella (pim-pih-NEL-uh). Anise.
Pinna (PIN-uh). Primary division of pinnate leaf.
Pinnata (pin-NAY-tuh). Pinnate, with leaflets on each side of a common petiole.
Pinus (PY-nus). Pine.
Piper (PY-pur). Pepper, cubeb.
Piperita (py-pur-EYE-tuh). Peppermint-scented.
Piqueria (py-KWEE-ree-uh). Stevia of florists.
Pisifera (py-SIF-ur-uh). Pea-bearing.
Pistacia (pih-STAY-she-uh). Pistachio tree.
Pistia (PIH-stee-uh). Water lettuce.
Pistillate (PIH-stih-layt). With pistils, without stamens.
Pisum (PY-sum). Garden pea.
Pittosporum (pih-TOSS-po-rum; pit-o-SPOH-rum). Australian laurel.
Pityrogramma (pit-ih-ro-GRAM-uh). Gold and silver ferns.
Placatus (play-KAY-tus). Quiet, calm.
Planera (PLAN-ur-uh; pla-NEE-ruh). Planer tree, water elm.
Plantago (plan-TAY-go). Plantain.
Platanus (PLAT-uh-nus). Plane tree, sycamore.
Platycentra (plat-ih-SEN-truh). With a broad center.
Platycerium (plat-ih-SEE-ree-um). Staghorn fern.

Platycodon (plat-ih-KO-don). Balloonflower.
Platyphylla (plat-ih-FIL-uh). With broad, flat leaves.
Platystemon (plat-ih-STEE-mon). Creamcups.
Plena (PLEE-nuh). Double, full.
Plicata (ply-KAY-tuh). Plaited, folded like a fan.
Plumarius (plyoo-MAY-ree-us). Plumed, feathered.
Plumbaginoides (plum-bah-jih-NOY-deez). Resembling plumbago.
Plumbago (plum-BAY-go). Leadwort.
Plumeria (plyoo-MEE-ree-uh). Frangipani, temple tree.
Plumosus (plyoo-MO-sus). Feathery.
Poa (PO-uh). Bluegrass, meadow grass.
Podocarpa (po-doh-KAHR-puh). With stalked fruits.
Podocarpus (pod-o-KAHR-pus; po-doh-KAHR-pus). Evergreen trees, shrubs.
Podophyllum (pod-o-FIL-um; po-doh-FIL-um). May apple, mandrake.
Poeticus (po-ET-ih-kus). Pertaining to poets.
Pogonia (po-GO-nee-uh). Bog orchids.
Poinciana (poyn-see-AY-nuh). Barbados pride, flower-fence; showy trees, shrubs.
Poinsettia (poyn-SET-ee-uh). Showy Christmas plants.
Polemonium (po-leh-MO-nee-um). Jacob's-ladder.
Polianthes (pah-lee-AN-theez). Tuberose.
Polyacanthus (pah-lee-uh-KAN-thus). Many-spined.
Polyantha (pah-lee-AN-thuh). With many flowers; kind of rose.
Polyanthus (pah-lee-AN-thus). Common name of hardy primulas.
Polygala (po-LIG-uh-luh). Milkwort, senega snakeroot.
Polygamous (po-LIG-uh-mus). Bearing perfect, staminate and pistillate flowers.
Polygonatum (po-lig-o-NAY-tum). Solomon's-seal.
Polygonum (po-LIG-o-num). Silver-lace vine, smartweed.
Polymorphus (pah-lee-MOR-fus). Having many forms.
Polyphyllus (pah-lee-FIL-us). With many leaves.

Polypodium (pah-lee-PO-dee-um). Huge genus of ferns; polypody.

Polypody (PAH-lih-po-dee). Tropical and hardy ferns.

Polypogon (pah-lee-PO-gon). Annual beard grass.

Polystachya (pah-lee-STAK-ee-uh). With many spikes.

Polystichum (po-LISS-tih-kum). Christmas dagger fern.

Pomaceus (po-MAY-see-us; po-MAY-shus). Like a pome.

Pome (POHM). Fleshy fruit, like apple.

Pomelo (POM-eh-lo). Another name for grapefruit.

Pomifera (po-MIF-ur-uh). Pome-bearing.

Pomology (po-MOL-o-jee). Study and culture of fruits.

Ponchirus (pon-SY-rus). Trifoliate or hardy orange.

Ponderosus (pon-dur-O-sus). Heavy, massive.

Pontederia (pon-tee-DEE-ree-uh). Pickerelweed.

Ponticus (PON-tih-kus). Of Pontus in Asia Minor.

Populus (POP-yoo-lus). Poplar, aspen, cottonwood.

Portulaca (por-tyoo-LAY-kuh; por-tyoo-LAK-uh as common name). Rose moss, purslane.

Portulacaria (por-tyoo-luh-KAY-ree-uh). Purslane tree.

Potentilla (po-ten-TIL-uh). Cinquefoil.

Poterium (po-TEE-ree-um). Japanese burnet.

Pothos (PO-thos). Climbing plants.

Praealtus (pre-AL-tus). Very tall.

Praecox (PRE-koks). Precocious, very early.

Praestans (PRE-stanz). Distinguished, excelling.

Pratensis (pray-TEN-siss). Of meadows.

Pratia (PRAY-tee-uh). Creeping rockery plants.

Prenanthes (pre-NAN-theez). With drooping flowers.

Primula (PRIM-yoo-luh). Primrose.

Primulinus (prim-yoo-LY-nus). Primrose-like.

Princeps (PRIN-seps). Princely, first.

Prinsepia (prin-SEE-pee-uh). Spiny, hardy Asiatic shrubs.

Pritchardia (prih-CHAR-dee-uh). Ornamental fan palms.

Privet (PRIH-vet). Common hedge plants; *Ligustrum*.

Proboscidea (pro-bo-SID-ee-uh). Unicorn plant.

Procerus (pro-SEER-us). Tall.
Procumbens (pro-KUM-benz). Trailing, prostrate.
Prolifera (pro-LIF-fer-uh). Many-leaved.
Prolificum (pro-LIF-ih-kum). Prolific.
Prosopis (pro-SO-piss). Mesquite.
Prostratus (pross-TRAY-tus). Lying flat.
Pruinosus (proo-ih-NO-sus). Frosted.
Prunella (proo-NEL-uh). Selfheal, heal-all.
Prunifolium (proo-nih-FO-lee-um). Plum-leaved.
Prunus (PROO-nus). Plum, cherry.
Pseudolarix (soo-doh-LAHR-iks; soo-doh-LAY-riks). Golden larch.
Pseudotsuga (soo-doh-SOO-guh). Douglas fir.
Psidium (SID-ee-um). Guava.
Psittacinus (sit-uh-SY-nus; sih-TAS-ih-nus). Parrot-like or parrot-colored.
Ptarmica (TAHR-mih-kuh). Sneeze-producing.
Ptarmicoides (tahr-mih-KOY-deez). Resembling sneezewort.
Ptelea (TEE-lee-uh). Hop tree, wafer ash.
Pteridium (teh-RID-ee-um). Brake, bracken.
Pteris (TEE-riss; TER-iss). Common table or dish ferns.
Pterostyrax (teer-ro-STY-raks). Asiatic shrubs, trees; ornamentals.
Ptychosperma (ty-ko-SPUR-muh). Slender feather palms.
Puberulus (pyoo-BUR-yoo-lus). Somewhat pubescent.
Pubescens (pyoo-BESS-enz). Pubescent, hairy—short, soft and downy.
Pudicus (PYOO-dih-kus). Shrinking.
Pueraria (pyoo-eh-RAY-ree-uh). Kudzu vine.
Pulchellus (puhl-KEL-us). Beautiful.
Pulcherrimus (puhl-KER-rih-mus). Very handsome.
Pulmonaria (puhl-mo-NAY-ree-uh). Lungwort, Bethlehem sage.
Pulverulentus (puhl-vur-yoo-LEN-tus). Powdered, dust-covered.
Pumilus (PYOO-mih-lus). Dwarf.
Punctata (punk-TAY-tuh). Dotted with depressions, spotted.

Pungens (PUN-jenz). Piercing, sharp-pointed, sharp tasting.
Punica (PYOO-nih-kuh). Pomegranate.
Puniceus (pyoo-NISS-ee-us). Reddish-purple.
Purpurascens (pur-pur-ASS-enz). Becoming purple.
Purpurea (pur-PYOO-ree-uh). Purple.
Puschkinia (poosh-KIN-ee-uh). Spring-blooming bulbs; from proper name.
Pusillus (pyoo-SIL-us). Very small.
Pustulatus (pus-tyoo-LAY-tus). As though blistered.
Puya (pyoo-yuh). Spiny desert plants.
Pychnostachys (pik-NOSS-tah-kiss). Stout perennial labiates.
Pycnanthemum (pik-NAN-theh-mum). With dense flowers.
Pygmaeus (pig-MEE-us). Pigmy, dwarf.
Pyracantha (py-ruh-KAN-thuh). Fire thorn.
Pyramidalis (pih-ram-ih-DAY-liss). Like a pyramid.
Pyrethrum (py-REE-thrum). Painted daisy.
Pyrola (PIR-o-luh). Wintergreen, shinleaf.
Pyrostegia (py-ro-STEE-jee-uh). Southern climbers.
Pyrus (PY-rus). Pear.

Q

Quamoclit (KWAM-o-klit). Cypress vine, cardinal climber.
Quassia (KWOSH-ee-uh). Tender medicinal trees.
Quercus (KWUR-kus). Oak.
Quillaja (kwih-LAY-yuh; kwih-LAY-juh). Soapbark tree.
Quinatus (kwih-NAY-tus). In fives.
Quinquefolia (kwin-kweh-FO-lee-uh). Five leaves or leaflets.

R

Raceme (ruh-SEEM; ray-SEEM). Pediceled flowers along one stem.
Racemosus (ra-seh-MO-sus; ra-see-MO-sus). Flowers in racemes.
Rachis (RAY-kiss). Center stem of spike or compound leaf.

Radiatus (ray-dee-AY-tus). Rayed.
Radicans (RAD-ih-kanz). Rooting, especially along stem.
Raffia (RAF-ee-uh). Fiber from raffia palm for tying plants.
Ramonda (ray-MON-duh). Low rockery herbs.
Ranunculus (ruh-NUNG-kyoo-lus). Buttercup, crowfoot.
Raoulia (ray-OO-lee-uh). Tufted or creeping rockery plants.
Raphanus (RAF-uh-nus). Radish.
Raphia (RAY-fee-uh; RAF-ee-uh). Raffia palm.
Raphiolepis (raf-ee-OL-eh-piss). India and Yeddo hawthorn.
Ratibida (ra-TIH-bih-duh; ra-tih-BIH-duh). Prairie coneflower.
Ravenala (rav-eh-NAY-luh). Traveler's tree of Madagascar.
Recurvata (reh-kur-VAY-tuh). Bent backward, reversely curved.
Recta (REK-tuh). Straight, erect.
Redivivus (reh-dih-VY-vus). Restored, brought to life.
Refractus (ree-FRAK-tus). Broken.
Refulgens (ree-FUL-jenz). Brightly shining.
Regalis (ree-GAY-liss). Regal, royal.
Regia (REE-jee-uh). Royal.
Reginae (ree-JEE-nuh). Queenly.
Rehmannia (ray-MAN-ee-uh). Showy, sticky perennials.
Reineckia (ry-NEK-ee-uh). Asiatic plants of lily family.
Reinwardtia (ryn-WAHR-tee-uh). Yellow flax.
Religiosus (ree-lij-ee-O-sus). Used for religious purposes.
Remotus (ree-MO-tus). With parts distant.
Renanthera (reh-nan-THEE-ruh; reh-NAN-thur-uh). Indo-Malayan orchids.
Reniformis (ren-ih-FORM-iss). Kidney-shaped.
Repens (REE-penz). Creeping.
Reptans (REP-tanz). Creeping.
Reseda (reh-SEE-duh). Mignonette.
Resiniferous (rez-in-IF-ur-us). Producing resin.
Resinosa (rez-ih-NO-suh). Resinous.
Reticulata (reh-tik-yoo-LAY-tuh). Netted-veined.
Retinispora (ret-ih-NISS-por-uh). Cypress-like evergreens.
Retusa (ree-TUS-uh). Notched.

Rex (REKS). King.
Rhamnus (RAM-nus). Buckthorn.
Rhapidophyllum (rap-ih-doh-FIL-um). Needle palm, blue palmetto.
Rhapis (RAY-piss). Low, reed-like, tufted fan palms.
Rheum (REE-um). Rhubarb, pieplant.
Rhexia (REKS-see-uh). Meadow beauty, deer grass.
Rhipsalis (RIP-suh-liss). Willow or mistletoe cactus.
Rhizoctonia (ry-zok-TOH-nee-uh). Plant disease fungus.
Rhizophora (ry-ZOF-o-ruh). Mangrove.
Rhodanthus (ro-DAN-thus). Flowers like roses.
Rhododendron (ro-doh-DEN-dron). Showy, flowering, mostly broad-leaved evergreen shrubs.
Rhodora (ro-DOH-ruh). Azalea-like shrubs.
Rhodotypos (ro-doh-TY-pos). Jetbead, white kerria.
Rhoeo (REE-o). Plant like wandering Jew; oyster plant.
Rhizome (RY-zohm). Prostrate or underground stem.
Rhizophyllus (ry-zoh-FIL-us). Leaves rooting, basal-leaved.
Rhus (RUS). Sumac.
Rhytidophyllus (ry-tih-doh-FIL-us; rit-ih-doh-FIL-us). Wrinkle-leaved.
Ribes (RY-beez). Currant, gooseberry.
Riccia (RIK-see-uh). Duckweed; floating water plants.
Richardia (rih-CHAR-dee-uh). Mexican clover; formerly calla.
Ricinus (RISS-ih-nus). Castor bean.
Rigens (RY-jenz). Rigid, stiff.
Rigidus (RIJ-ih-dus). Stiff.
Ringens (RIN-jenz). Gaping.
Riparius (ry-PAYR-ee-us). Of river banks.
Rivalis (rih-VAY-liss). Pertaining to brooks.
Rivina (rih-VY-nuh). Rouge plant, bloodberry.
Rivularis (riv-yoo-LAY-riss). Brook-loving.
Robinia (ro-BIN-ee-uh). Locust, rose acacia.
Robur (RO-bur). Species name of English oak.
Robustus (ro-BUS-tus). Stout, vigorous.

Rochea (RO-kee-uh). Red-flowered florists' plant.

Rodgersia (rod-JUR-see-uh). Hardy, robust perennials; from proper name.

Rohdea (RO-dee-uh). Durable foliage plant.

Romneya (ROM-nee-uh; rom-NEE-yuh). Matilija or giant poppy.

Rondeletia (ron-deh-LEE-she-uh). Tropical evergreen shrubs.

Rosa (RO-zuh). Rose.

Rosaceae (ro-ZAY-see-ee). Rose family.

Rosa-sinensis (ro-zuh-sih-NEN-siss). Rose of China.

Roseus (RO-zee-us). Rosy, rose-colored.

Rosmarinus (ross-muh-RY-nus; roz-muh-RY-nus). Rosemary.

Rostrata (ross-TRAY-tuh). Beaked.

Rotundifolius (ro-tun-dih-FO-lee-us). Round-leaved.

Roystonea (roy-STO-nee-uh). Royal and cabbage palms.

Ruber (ROO-bur). Red, ruddy.

Rubescens (roo-BESS-enz). Becoming red.

Rubia (ROO-bee-uh). Madder.

Rubiginosa (ruh-big-ih-NO-suh). Rusty.

Rubra (ROO-bruh). Red.

Rubrofructa (roo-bro-FRUK-tuh). Red-fruited.

Rubus (ROO-bus). Brambles; raspberry, blackberry, dewberry.

Rudbeckia (rud-BEK-ee-uh). Coneflower.

Ruellia (roo-EL-ee-uh). Manyroot.

Rufescens (roo-FESS-enz). Becoming red.

Rufidulus (roo-FID-yoo-lus). Somewhat reddish.

Rugosa (roo-GO-suh). Wrinkled.

Rumex (ROO-meks). Dock, sorrel.

Rupestris (roo-PESS-triss). Rock-loving.

Rupicolus (roo-PIK-o-lus). Growing on cliffs or ledges.

Ruscus (RUS-kus). Butcher's-broom.

Russelia (ruh-SEE-lee-uh; ruh-SEEL-yuh). Coral or fountain plant.

Rusticanus (rus-tih-KAY-nus). Pertaining to the country.

Ruta (ROO-tuh). Rue.

Rutilans (ROO-tih-lanz). Red or becoming red.

S

Sabal (SAY-bal). Palmetto; common fan palm.
Sabatia (sa-BAY-she-uh). American centaury, sea pink.
Saccatus (sa-KAY-tus). Bag-like.
Saccharatus (sak-uh-RAY-tus). Containing sugar, sweet.
Saccharum (sak-KAHR-um). Sugar; also genus of sugar cane, species of maple.
Sachalinensis (sa-kuh-len-EN-siss). Of Saghalien Island (Southeastern Russia).
Sacrorum (sa-KRO-rum). Sacred, of sacred places.
Sagina (suh-JY-nuh). Pearlwort, pearlweed.
Sagittaria (saj-ih-TAY-ree-uh). Arrowhead.
Saguaro (suh-WAH-roh). Giant cactus.
Saintpaulia (saynt-PAW-lee-uh). African violet.
Salicaria (sal-ih-KAY-ree-uh). Like a willow.
Salicifolia (sal-iss-ih-FO-lee-uh). Leaves like willow.
Salinus (suh-LY-nus). Salty, of salty places.
Salix (SAY-liks). Willow, osier.
Salpiglossis (sal-pih-GLOSS-iss). Painted tongue.
Salvia (SAL-vee-uh). Scarlet sage, sage, silver sage.
Samara (SAM-uh-ruh; suh-MAY-ruh). Winged, with one-seeded fruits.
Sambucus (sam-BYOO-kus). Elder.
Sandersoni (san-dur-SO-nye). From proper name.
Sanguinaria (sang-gwih-NAY-ree-uh). Bloodroot.
Sanguineus (san-GWIN-ee-us). Bloody, blood-red.
Sanguisorba (sang-gwih-SOR-buh). Burnet.
Sansevieria (san-seh-vih-EE-ree-uh). Bowstring hemp.
Santalum (SAN-tuh-lum). Sandalwood.
Santolina (san-toh-LY-nuh). Lavender cotton, aromatic undershrubs.
Sanvitalia (san-vih-TAY-lee-uh). Creeping zinnia.
Sapidus (SAP-ih-dus). Savory, pleasing to taste.
Sapientum (say-pee-EN-tum). Of wise men or authors.

Sapindus (suh-PIN-dus). Soapberry.
Sapium (SAY-pee-um). Chinese tallow tree.
Saponaria (sap-o-NAY-ree-uh). Soapwort.
Sapota (suh-PO-tuh). Sapodilla, sapote.
Sarcococca (sar-ko-KOK-uh). Asiatic, Malayan, low-growing, evergreen shrubs resembling box.
Sarmentosus (sahr-men-TOH-sus). Bearing runners.
Sarracenia (sar-uh-SEE-nee-uh). Pitcher plant.
Sasa (SAH-sah). From Japanese name for small bamboos.
Sasanqua (sa-SAN-kwah). Plum-flowered tea.
Sassafras (SASS-uh-frass). Aromatic trees, often shrubby.
Sativus (sa-TY-vus). Cultivated.
Satureja (sat-yoo-REE-yuh). Savory.
Saxatilis (saks-AT-ih-liss). Found among rocks.
Saxicolus (saks-ih-KO-lus). Rock dweller.
Saxifraga (saks-IF-ruh-guh). Saxifrage, rockery plants.
Scabiosa (skay-bee-O-suh). Scabious, pincushion flower.
Scabrous (SKAY-brus). Rough to the touch.
Scandens (SKAN-denz). Climbing.
Scape (SKAYP). Leafless flower stem from ground.
Scariosus (skayr-ee-O-sus). Thin and not green.
Schefflera (shef-LEER-uh). Shrubs, trees of Araliaceae.
Schinus (SKY-nus). California pepper tree.
Schisandra (sky-ZAN-druh; skih-ZAN-druh). Aromatic woody vines.
Schizachyrium (skih-ZAK-ree-um; skih-zah-KY-ree-um). Split chaff.
Schizanthus (sky-ZAN-thus; skih-ZAN-thus). Butterfly flower.
Schizopetalus (skih-zo-PET-uh-lus). Cut-petaled.
Schizophragma (skih-zo-FRAG-muh). Japanese hydrangea vine.
Schomburgkia (shom-BUR-kee-uh). Showy orchids.
Sciadopitys (sy-uh-DOP-ih-tiss). Umbrella pine.
Scilla (SIL-uh). Squill, Spanish bluebell, Cuban lily.
Scilloides (sil-OY-deez). Squill-like.
Scindapsus (sin-DAP-sus). Pothos of florists.

Scirpus (SUR-pus). Bulrush, club rush.
Scleria (SKLEE-ree-uh). Nut grass.
Sclerocactus (skleer-o-KAK-tus; skler-o-KAK-tus). Globular cactus.
Scoparia (sko-PAY-ree-uh). Broom or broom-like.
Scopulorum (skop-yoo-LOR-um). Of the rocks.
Scrophularia (skrof-yoo-LAY-ree-uh). Figwort.
Scutellaria (skyoo-teh-LAY-ree-uh). Skullcap.
Scutum (skyoo-tum). A shield.
Sechium (SEE-kee-um). Chayote.
Secundus (see-KUN-dus). Side-flowering.
Sedum (SEE-dum). Stonecrop; rockery succulents.
Segetum (SEG-eh-tum). Of cornfields.
Selaginella (sel-uh-jih-NEL-uh). Moss-like and fern-like plants.
Selenicereus (seh-lee-ni-SEE-ree-us). Night-blooming cactus.
Selenipedium (seh-lee-nih-PEE-dee-um). Tropical orchids.
Semperflorens (sem-pur-FLO-renz). Everflowering.
Sempervirens (sem-pur-VY-renz). Evergreen.
Sempervivum (sem-pur-VY-vum). Houseleek.
Senecio (seh-NEE-she-oh). Florists' cineraria, groundsel.
Senilis (seh-NY-liss). Old, white-haired.
Sensitivus (sen-sih-TY-vus). Sensitive.
Sepal (SEE-pal; SEP-al). Division of a calyx.
Septemlobus (sep-tem-LO-bus). Seven-lobed.
Sequoia (see-KWOY-uh). Redwood, big tree.
Serenoa (ser-eh-NO-uh). Scrub or saw palmetto.
Sericeus (ser-ISS-ee-us). Silky.
Serotinus (ser-OT-ih-nus). Late, late-flowering, late-ripening.
Serpens (SUR-penz). Creeping, crawling.
Serrata (ser-RAY-tuh). Serrate, toothed, teeth pointing forward.
Serrulate (SER-yoo-layt). Finely saw-toothed.
Sesamum (SESS-uh-mum). Sesame.
Sessifolius (sess-ih-FO-lee-us). Stalkless leaves.
Sessile (SESS-il). Without a stalk.
Setaceus (seh-TAY-see-us; see-TAY-see-us). Bristle-like.

Setaria (seh-TAY-ree-uh). Foxtail millet, palm grass.
Setigera (seh-TIJ-ur-uh; see-TIJ-ur-uh). Bristle-bearing.
Setose (SEE-tohs; see-TOHS). Beset with bristles.
Shepherdia (sheh-PUR-dee-uh; shep-HUR-dee-uh). Buffalo berry.
Shortia (SHOR-tee-uh). Low, evergreen herbs.
Sibiricus (sy-BIR-ih-kus). Of Siberia.
Sidalcea (sy-DAL-she-uh; sih-DAL-she-uh). Mallow-like herbs.
Signatus (sig-NAY-tus). Marked, designated.
Silene (sy-LEE-nee). Catchfly, campion, cushion pink.
Silique (sih-LEEK; SIL-ik). Seed pod of mustard, cabbage.
Silky (SIL-kee). Having close-pressed, soft, straight hairs.
Silphium (SIL-fee-um). Compass plant, Indian cup.
Similis (SIM-ih-liss). Similar, like.
Simplex (SIM-pleks). Simple.
Sinensis (sy-NEN-siss). Of China.
Sinningia (sih-NIN-jee-uh). Gloxinia.
Sinuata (sin-yoo-AY-tuh). Margin strongly wavy.
Sisyrinchium (siss-ih-RING-kee-um). Blue-eyed grass.
Skimmia (SKIM-ee-uh). Asiatic evergreen shrubs.
Smilacina (smy-luh-SY-nuh). False Solomon's-seal.
Smilax (SMY-laks). Greenbrier, cat brier.
Solandra (so-LAN-druh). Chalice vine.
Solanum (so-LAY-num). Nightshade, Jerusalem cherry, potato.
Soldanella (sol-duh-NEL-uh). Alpine herbs.
Solidago (sol-ih-DAY-go). Goldenrod.
Sollya (SOL-ee-uh; SOL-yuh). Australian bluebell creeper.
Soongarica (son-GAYR-ih-kuh; soon-GAYR-ih-kuh). Of Songchin, Korea.
Sophora (so-FO-ruh; SOF-o-ruh). Pagoda or Chinese scholar tree.
Sophronitis (so-fro-NY-tiss). Brazilian orchids.
Sorbaria (sor-BAY-ree-uh). False spirea.
Sorbus (SOR-bus). Mountain ash.
Sorghastrum (sor-GAS-trum). Perennial grass.

Sorus (SO-rus). Spore case on ferns.
Spadix (SPAY-diks). Spike with fleshy axis, as in calla.
Sparaxis (spuh-RAK-siss). Wandflower.
Spartina (spar-TEE-nuh; spar-TY-nuh). Cord-like.
Spartium (SPAHR-she-um). Spanish broom.
Spathe (SPAYTH). Large bract enclosing spadix, as in calla.
Spathyphyllum (spath-ih-FIL-um). Showy aroid.
Spathulatus (spath-yoo-LAY-tus). Spoon-shaped.
Species (SPEE-sheez). Subdivision of genus, second scientific name.
Speciosus (spee-see-O-sus). Showy, good-looking.
Spectabilis (spek-TAB-ih-liss). Remarkable, showy.
Specularia (spek-yoo-LAY-ree-uh). Venus's-looking glass.
Sphaeralcea (sfee-RAL-see-uh). Globe mallow.
Sphaericus (SFEER-ih-kus). Spherical.
Sphaerocephala (sfeer-o-SEF-uh-luh). Round headed.
Sphagnum (SFAG-num). A bog moss.
Spicatus (spy-KAY-tus). Spicate, with spikes.
Spigelia (spy-JEE-lee-uh). Pinkroot.
Spike (SPYK). Flowers sessile on common stem.
Spinacia (spih-NAY-she-uh). Spinach.
Spinosissimus (spy-no-SISS-ih-mus). Most or very spiny.
Spinosus (spy-NO-sus). Full of spines.
Spiraea (spy-REE-uh). Hardy, flowering shrubs; spirea.
Spiralis (spy-RAY-liss; spy-RAL-iss). Spiral.
Spiranthes (spy-RAN-theez). Terrestrial orchids, lady's-tresses.
Splendens (SPLEN-denz). Splendid.
Spodiopogon (spo-dee-o-POH-gon). Grey-bearded.
Spondias (SPON-dee-as). Otaheite apple, hog plum, mombin.
Sporobolus (spor-uh-BOH-lus). Dropseed.
Sprekelia (spreh-KEE-lee-uh). Jacobaean or St.-James's-lily.
Sprengeri (SPRENG-ur-eye). From proper name.
Spruce (SPROOS). *Picea*; tall coniferous trees.
Spurius (SPYOO-ree-us). Spurious, false.
Squamatus (skwah-MAY-tus). With scale-like leaves or bracts.

Squarrosus (skwah-RO-sus). With parts spreading at ends.
Stachys (STAY-kiss). Lamb's-ears, betony.
Stamen (STAY-men). Male organ in flowers; bears pollen.
Staminate (STAM-ih-nayt). Having stamens and no pistil.
Stanhopea (stan-HO-pee-uh). Tropical American orchids.
Stapelia (stuh-PEE-lee-uh). Carrion or starfish flower.
Staphylea (staf-ih-LEE-uh). Bladdernut.
Statice (STAT-ih-see). Thrift, sea pink.
Stauntonia (stawn-TOH-nee-uh). Woody Asiatic vines.
Steironema (sty-ro-NEE-muh). Loosestrife.
Stellaria (steh-LAY-ree-uh). Starwort, Easter bells.
Stellata (steh-LAY-tuh). Star-like.
Stenocephalus (sten-o-SEF-uh-lus). Narrow-headed.
Stenolobium (sten-o-LO-bee-um). Yellow elder.
Stenopetalus (sten-o-PET-uh-lus). Narrow-petaled.
Stenophyllus (sten-o-FIL-us). Narrow-leaved.
Stenotaphrum (sten-o-TAF-rum). St. Augustine grass.
Stephanandra (stef-uh-NAN-druh). Hardy Asiatic shrubs.
Stephanotis (stef-uh-NO-tiss). Madagascar jasmine, waxflower.
Sterile (STER-il). Unproductive, not fertile.
Sternbergia (sturn-BUR-jee-uh). Winter daffodil.
Stevia (stee-vee-uh). Florists' filler flower; piqueria.
Stewartia (styoo-AHR-tee-uh; styoo-AHR-she-uh). Mountain camellia.
Stigma (STIG-muh). Sticky end of pistil.
Stipa (STY-puh). Feather, silk or bunch grass.
Stipule (STIP-yool). Bract-like appendage at base of leafstalk.
Stokesia (sto-KEE-zhee-uh; sto-KEE-see-uh). Stokes' aster.
Stolon (STO-lon). Runner, basal branch.
Stoloniferous (sto-lon-IF-ur-us). Producing runners that root.
Stranvaesia (stran-VEE-zee-uh). Southern evergreen shrubs.
Strelitzia (streh-LIT-see-uh). Bird-of-paradise flower.
Streptocarpus (strep-tow-KAHR-pus). Cape primrose; also twisted fruit.
Striata (stry-AY-tuh). Striped.

Stricta (STRIK-tuh). Upright, with few or no branches.
Strigose (STRY-gohs; strih-GOHS). With appressed sharp, stiff hairs.
Strobilanthes (stro-bih-LAN-theez). Hothouse foliage plants.
Strobile (STROB-il). Inflorescence marked by scales, as in hop and pine cone.
Strycnos (STRIK-nos). Tropical trees, vines; source of strychnine.
Style (STYL). Stem part of pistil.
Stylophorum (sty-LOFF-o-rum). Celandine poppy.
Stylosus (sty-LO-sus). With prominent styles.
Styraciflua (sty-rah-seh-FLOO-uh). Flowing with storax or gum.
Styrax (STY-raks). Snowbell, storax.
Suaveolens (swah-vee-O-lenz). Sweet-scented.
Subalpinus (sub-al-PY-nus). Nearly alpine.
Subcaeruleus (sub-seh-ROO-lee-us; sub-keh-ROO-lee-us). Slightly blue.
Subcarnosus (sub-kahr-NO-sus). Rather fleshy.
Subcordatus (sub-kor-DAY-tus). Almost heart-shaped.
Subsessilis (sub-SESS-ih-liss). Nearly stalkless.
Subulatus (sub-yoo-LAY-tus). Awl-shaped.
Suecicus (SWEE-sih-kus). Of Sweden.
Suffrutescent (suf-roo-TESS-ent). Slightly shrubby.
Suffruticosa (suh-froo-tih-KO-suh). Somewhat shrubby.
Sulphureus (sul-FYOO-ree-us). Yellow.
Superba (syoo-PUR-buh). Showy, superb, proud.
Supinus (syoo-PY-nus). Prostrate.
Susianus (syoo-see-AY-nus). Of Susa, ancient Persian city.
Suspensa (sus-PEN-suh). Drooping.
Suwarro (suh-WAH-ro). Giant cactus; carnegiea.
Swainsona (swayn-SO-nuh). Flowering greenhouse plant.
Swietenia (swee-TEE-nee-uh). Mahogany.
Syagrus (sy-AY-grus). Palm known as cocos in trade.
Sylvaticus (sil-VAT-ih-kus). Forest-loving.
Sylvestris (sil-VESS-triss). Of woods or forests.

Symphoricarpos (sim-for-ih-KAHR-pos). Snowberry, Indian currant.
Symphyandra (sim-fee-AN-druh). Campanula-like plants.
Symphytum (sim-FY-tum). Comfrey.
Symplocarpus (sim-plo-KAHR-pus). Skunk cabbage.
Symplocos (sim-PLO-kus). Sweetleaf.
Syncarpia (sin-KAHR-pee-uh). Turpentine tree.
Syriacus (seer-ee-AY-kus). Of Syria.
Syringa (sih-RING-guh). Lilac.
Syzygium (sy-ZY-gee-um). Coupled, paired.

T

Tabebuia (tab-beh-BOO-yuh). Ornamental trees; from Brazilian name.
Tabernaemontana (tuh-bur-nee-mon-TAY-nuh). Crape jasmine.
Tabuliformis (tab-yoo-lih-FOR-miss). Table-like.
Tagetes (tuh-JEE-teez). Marigold.
Talinum (tuh-LY-num). Rock pink.
Tamarindus (tam-uh-RIN-dus). Tamarind.
Tamariscifolia (tam-uh-riss-sih-FO-lee-uh). Leaves like tamarix.
Tamarix (TAM-uh-riks). Tamarisk.
Tamus (TAY-mus). Black bryony.
Tanacetum (tan-uh-SEE-tum). Tansy.
Tanguticus (tan-GYOO-tih-kus). Of or near Tangut, Tibet.
Taraxacum (tuh-RAK-suh-kum). Dandelion.
Tartaricus (tuh-TAR-ih-kus). Of Tatary, or central Asia.
Taxaceae (taks-AY-see-ee). *Taxus* or yew family.
Taxifolius (taks-ih-FO-lee-us). Yew-leaved.
Taxodium (taks-O-dee-um). Bald and Montezuma cypress.
Taxus (TAK-sus). Yew.
Tecoma (teh-KO-muh). Tender Southern shrubs.
Tecomaria (tek-o-MAY-ree-uh). Cape honeysuckle.
Tectona (tek-TOH-nuh; TEK-toh-nuh). Teak.
Tectorum (tek-TOH-rum). Of roofs or houses.

Telanthera (tel-AN-thur-uh). Carpet bedding plants.
Telephium (teh-LEE-fee-um). Orpine.
Tellima (teh-LY-muh). False alumroot, fringe cup.
Tenax (TEE-naks). Tenacious, strong.
Tenellus (teh-NEL-us). Tender, delicate.
Tenera (TEN-ur-uh). Slender, tender, soft.
Tenuifolius (ten-yoo-ih-FO-lee-us). Slender-leaved.
Tenuis (TEN-yoo-iss). Slender, thin.
Tephrosia (teh-FRO-zhee-uh; tef-RO-see-uh). Goat's-rue, catgut.
Terete (TER-eet). Cylindrical.
Terminalis (tur-mih-NAY-liss; Tur-min-AWL-iss). At the end.
Ternatus (tur-NAY-tus). In threes, as clover leaflets.
Ternstroemia (turn-STREE-mee-uh). From proper name.
Tessellata (tess-eh-LAY-tuh). Checkered.
Testaceus (tess-TAY-see-us). Light brown, dull brick-red.
Tetragonia (tet-ruh-GO-nee-uh). New Zealand spinach.
Tetragonus (tet-ruh-GO-nus). Four-angled.
Tetrapanax (teh-TRAP-uh-naks). Rice-paper tree.
Tetraptera (teh-TRAP-tur-uh). Four-winged.
Tetrastigma (tet-rah-STIG-muh). Four-lobed stigma.
Teucrium (TYOO-kree-um). Germander.
Texanus (tek-SAY-nus). *Texensis*, of Texas.
Textilis (teks-TIL-iss). Woven, for weaving.
Thelia (THAY-lee-uh). Water canna.
Thalictrifolia (thuh-lik-trih-FO-lee-uh). Leaves like *Thalictrum*.
Thalictrum (thuh-LIK-trum). Meadow rue.
Thallus (THAL-us). Body of mosses, lichens, algae.
Thea (THEE-uh). Tea.
Theifera (thee-IF-ur-uh). Tea-bearing.
Thelesperma (thel-eh-SPUR-muh). Coreopsis-like annuals.
Thelocactus (thel-o-KAK-tus). Globular cactus of Texas.
Thelypteris (theh-LIP-tur-iss). Tender fern.
Theobroma (thee-o-BRO-muh). Chocolate tree of tropics.
Thermopsis (thur-MOP-siss). Yellow pea-flowered perennials.
Thevetia (theh-VEE-she-uh). Yellow oleander.

Thlaspi (THLAS-py). Pennycress.
Thrinax (THRY-naks). Thatch palms.
Thryallis (thry-AL-iss). Handsome shrub sometimes forced.
Thuja (THOO-yuh). Arborvitae.
Thujopsis (thoo-YOP-siss). False or Hiba arborvitae.
Thunbergia (thun-BUR-jee-uh). Black-eyed Susan vine.
Thuringiaca (thur-in-GEE-ih-kuh). Of Thuringa, Germany.
Thymus (THY-mus). Thyme; rockery plants; old herbs.
Thyoides (thy-OY-deez; thee-OY-deez). Like thuja.
Thyrsoides (thur-SOY-deez). Thyrse-like.
Tiarella (ty-uh-REL-uh). Foamflower.
Tibouchina (tib-oo-KY-nuh). Spiderflower, glory bush.
Tigridia (ty-GRID-ee-uh). Tigerflower.
Tigrinus (tih-GRY-nus; ty-GRY-nus). Striped like a tiger.
Tilia (TIL-ee-uh). Linden or basswood tree.
Tillandsia (tih-LAND-zee-uh). Spanish or southern moss.
Tinctorius (tink-TOH-ree-us). For dyeing, used by dyers, bright.
Tingitanus (tin-jih-TAY-nus). Of Tangier, Morocco.
Titanus (ty-TAY-nus). Very large.
Tithonia (tih-THO-nee-uh). Mexican sunflower.
Tolmiea (tol-MEE-uh). Sticky, hairy perennials.
Tolpis (TOL-piss). Crepis-like annuals.
Tomentosa (toh-men-TOH-suh). Dense with soft hairs.
Topiary (TOH-pee-ay-ree). Art of trimming trees or shrubs.
Torenia (toh-REE-nee-uh). Low free-flowering annuals.
Toringoides (tor-in-GOY-deez). Like toringo crab.
Torreya (TOR-ee-uh; tor-EE-uh). California nutmeg.
Tortifolius (tor-tih-FO-lee-us). Leaves twisted.
Tortuosus (tor-tyoo-O-sus). Much twisted.
Townsendia (town-ZEN-dee-uh). Rockery plants; Easter daisy.
Toxicodendron (tok-sih-ko-DEN-dron). Old name for poison ivy.
Toxicus (TOKS-ih-kus). Poisonous.
Toyon (TOH-yun). California holly, Christmas berry.
Trachelium (truh-KEE-lee-um). Throatwort.

Trachelospermum (tray-kee-lo-SPUR-mum; truh-kel-o-SPUR-mum). Star jasmine.
Trachymene (truh-KY-meh-nee). Blue laceflower.
Tradescantia (trad-ess-KAN-she-uh). Wandering Jew, spiderwort.
Tragopogon (trag-o-PO-gon). Salsify, or vegetable oyster.
Trapa (TRAY-puh; TRAP-uh). Water chestnut.
Tremulus (TREM-yoo-lus). Quivering, trembling.
Triacanthophorus (try-uh-kan-THOF-o-rus). Bearing three spines.
Triacanthus (try-uh-KAN-thus). Three-spined.
Trianae (try-AY-nee). Common cattleya species.
Trichocarpus (try-ko-KAHR-pus; trih-ko-KAHR-pus). Hairy-fruited.
Trichosanthes (trik-o-SAN-theez). Snake or club gourd.
Trichosporum (try-KOS-por-um; trik-o-SPO-rum). Blushwort.
Trichostema (trik-o-STEE-muh). Bluecurls.
Tricolor (TRY-kul-ur). Three-colored.
Tricornis (try-KOR-niss). Three-horned.
Tricuspidatus (try-kus-pih-DAY-tus). Having three points.
Tricyrtis (try-KUR-tiss; try-SUR-tiss). With three humps; toad lily.
Trientalis (try-en-TAY-liss). Starflower.
Trifidus (TRIF-ih-dus). Three-parted.
Trifoliolate (try-FO-lee-o-layt). Having three leaflets.
Trifolium (try-FO-lee-um). Clover.
Trilisa (TRIL-ih-suh). Carolina or wild vanilla.
Trillium (TRIL-ee-um). Wakerobin.
Trilobata (try-lo-BAY-tuh). Three-lobed.
Triosteum (try-OS-tee-um). Horse gentian, feverwort.
Tripinnatifid (try-pih-NAT-ih-fid). Thrice pinnately cleft.
Trisetum (try-SEE-tum). False oats.
Tristania (trih-STAY-nee-uh; triss-TAHN-yuh). Evergreen flowering shrubs; from proper name.
Tristis (TRISS-tiss). Sad, bitter, dull.
Trithrinax (try-THRY-naks). Spiny fan palms.

Triticum (TRIT-ih-kum). Wheat.
Tritoma (try-TOH-muh). Torch lily, poker plant; now *Kniphofia*.
Tritonia (try-TOH-nee-uh). Gladiolus-like plants; montbretia.
Trivialis (trih-vee-AY-liss). Common, ordinary.
Trollius (TROL-ee-us). Globeflower.
Tropaeolum (troh-PEE-o-lum). Nasturtium.
Truncata (trung-KAY-tuh). Cut off squarely.
Tsuga (TSOO-guh). Hemlock.
Tuberosa (too-bur-O-suh). Bearing tubers.
Tulipa (TYOO-lih-puh). Tulip.
Tulipifera (tyoo-lih-PIF-ur-uh). Tulip-like.
Tunica (TYOO-nih-kuh). Tunic flower, saxifrage pink.
Tunicate (TYOO-nih-kayt). Concentric coats, as an onion.
Turbinata (tur-bih-NAY-tuh). Top-shaped.
Tupidanthus (tyoo-pih-DAN-thus). With hammer-shaped flowers.
Turgid (TUR-jid). Full of water, distended.
Typha (TY-fuh). Cattails.
Typhinus (ty-FY-nus). Pertaining to fever.

U

Uliginosus (yoo-lih-jih-NO-sus; yoo-lij-ih-NO-sus). Of wet or marshy places.
Ulmus (UL-mus). Elm.
Umbel (UM-bel). Flower cluster with stems springing from same point.
Umbellatus (um-bel-LAY-tus). With umbels.
Umbraculifera (um-brak-yoo-LIF-ur-uh). Umbrella-bearing.
Umbellularia (um-bel-yoo-LAYR-ee-uh). Umbrella-like.
Umbrosus (um-BROH-sus). Shade-loving.
Uncinatus (un-sih-NAY-tus). Hooked at the tip.
Undulate (UN-dyoo-layt). Wavy surface or margin.
Unguicularis (un-gwik-yoo-LAYR-iss). Clawed.
Unifolius (yoo-nih-FO-lee-us). Single-leaved.

Uniola (yoo-NEE-o-luh). Ornamental grass; sea oats.
Unisexual (yoo-nih-SEKS-yoo-al). Staminate or pistillate only.
Univittatus (yoo-nih-vih-TAY-tus). One-striped.
Urens (YOO-renz). Burning, stinging.
Urginea (ur-JIN-ee-uh). Sea onion; squill of medicine.
Ursinia (ur-SIN-ee-uh). Showy garden annuals.
Urtica (UR-tih-kuh; ur-TY-kuh). Nettle.
Urticifolius (ur-tih-kih-FO-lee-us). Nettle-leaved.
Utilis (YOO-tih-liss). Useful.
Utricularia (yoo-trik-yoo-LAY-ree-uh). Bladderwort.
Utriculatus (yoo-trik-yoo-LAY-tus). With bladdery, one-seeded fruit.
Uva-ursi (yoo-vuh-UR-see). Bear's grape; species of bearberry.
Uvularia (yoo-vyoo-LAY-ree-uh). Bellwort, wood daffodil.

V

Vaccinium (vak-SIN-ee-um). Blueberry, cranberry.
Vacillans (va-SIL-anz). Swaying.
Vagans (VAY-ganz). Vagrant, wandering.
Vaginalis (va-jih-NAY-liss). Sheathed.
Valeriana (vuh-lee-ree-AY-nuh). Valerian, garden heliotrope.
Valerianella (vuh-lee-ree-ay-NEL-uh). Italian corn salad.
Validus (VAL-ih-dus). Strong.
Vallisneria (val-iss-NEE-ree-uh). Eel or tape grass.
Vallota (va-LO-tuh). Scarborough lily.
Valvate (VAL-vayt). Opening by doors, or valves.
Valve (VALV). One of pieces into which a capsule splits.
Vancouveria (van-koo-VEE-ree-uh). Evergreen undershrubs.
Vanda (VAN-duh). Indo-Malayan tree-perching orchids.
Vanilla (vuh-NIL-uh). Climbing orchids, including vanilla vine.
Variabilis (vayr-ee-AB-ih-liss). Variable.
Varicosa (vayr-ih-KO-suh). Swollen irregularly.
Variegatus (vayr-ee-GAY-tus). Variegated.
Vascular (VASS-kyoo-lur). Furnished with ducts.

Vegetus (VEJ-eh-tus). Vigorous.
Veitchi (VEETCH-ee). From proper name, for nursery in England.
Velutinus (veh-LOO-tih-nus). Velvety.
Venidium (veh-NID-ee-um). South African daisies.
Venosus (veh-NO-sus). Veiny.
Ventricosum (ven-trih-KO-sum). Irregularly swollen.
Venustus (veh-NUS-tus). Handsome, charming.
Vera (VEE-ruh). True.
Veratrum (veh-RAY-trum). False and American hellebores.
Verbascum (vur-BASS-kum). Mullein.
Verbena (vur-BEE-nuh). Vervain, herbaceous plants.
Veris (VEE-riss; VER-iss). True.
Vernalis (vur-NAY-liss; vur-NAL-iss). Of spring.
Vernix (ver-NIKS). Varnish.
Vernonia (vur-NO-nee-uh). Ironweed.
Vernus (VUR-nus). Of spring.
Veronica (veh-RON-ih-kuh). Speedwell.
Verrucosa (ver-roo-KO-suh; ver-oo-KO-suh). Warty.
Verschaffelti (vur-shaf-FEL-tee). From proper name.
Versicolor (vur-SIK-o-lor). Variously colored, variegated.
Verticillata (vur-tih-sih-LAY-tuh). In circles around stem.
Verus (VEE-rus; VER-us). True, genuine or standard.
Vesicaria (ves-ih-KAY-ree-uh). Bladderpod.
Vestitus (ves-TY-tus). Clothed, covered with hairs, scales, etc.
Vexillaria (vek-sih-LAY-ree-uh). Pertaining to the standard petal in pea flower.
Viburnum (vy-BUR-num). Deciduous and evergreen woody shrubs.
Vicia (VISH-ee-uh; VISS-ee-uh). Vetch.
Victoria (vik-TOH-ree-uh). Royal water lily.
Vigna (VIG-nuh). Asparagus bean, cowpea.
Villosus (vil-LO-sus). Soft, hairy.
Vinca (VING-kuh). Periwinkle, creeping myrtle.
Vinifera (vy-NIFF-ur-uh; vih-NIFF-ur-uh). Wine-bearing.

Viola (VY-o-luh). Violet, pansy.
Violaceae (vy-o-LAY-see-ee). Violet family.
Virens (VY-renz). Green.
Virgatus (vur-GAY-tus). Twiggy.
Virginalis (vur-jih-NAY-liss; vur-jih-NAL-iss). White.
Virginiana (vur-jin-ee-AY-nuh). From Virginia.
Viridescens (vir-ih-DESS-senz). Becoming green.
Viridiflorus (vih-rih-dih-FLO-rus). Green-flowered.
Viridis (VIR-ih-diss). Green.
Viscidus (VISS-sih-dus). Viscid, sticky.
Viscosus (viss-KO-sus). Sticky.
Vitaceae (vy-TAY-see-ee). Grape or vine family.
Vitalba (vy-TAL-buh). White or white-flowered vine.
Vitellinus (vih-tel-LY-nus; vy-tel-LY-nus). Dull yellow approaching red.
Vitex (VY-teks). Chaste or hemp tree.
Vitis (VY-tiss). Woody vines; grape.
Vitis idaea (VY-tiss eye-DEE-uh). Specific name of mountain cranberry.
Vittatus (vih-TAY-tus). Striped.
Volubilis (vo-LYOO-bih-liss). Twining.
Vomitorius (vom-ih-TOH-ree-us). Emetic.
Vulgaris (vul-GAY-riss). Common.

W

Wahlenbergia (wah-len-BUR-jee-uh). New Zealand bluebell.
Waldsteinia (wald-STY-nee-uh). Barren strawberry.
Wallichia (wahl-LIK-ee-uh). Feather palm.
Washingtonia (wah-shing-TOH-nee-uh). Massive fan palm.
Washingtoniana (wah-shing-TOH-nee-AY-nuh; wah-shing-TOH-nee-A-nuh). Of Washington.
Watsonia (wot-SO-nee-uh). Gladiolus-like plants.
Weddeliana (wed-del-ee-AY-nuh). From proper name.
Weigela (wy-GEE-luh). Showy, hardy shrubs.

Whorl (HWURL; HWORL). Leaves, etc., in circle around stem.
Wichuraiana (wih-shur-ay-AN-uh). Given in honor of Wichuray, a Russian botanist.
Wilcoxia (wil-KOKS-ee-uh). Slender-stemmed cactus.
Wisetonensis (wyz-toh-NEN-siss). A form of *Schizanthus*.
Wistaria (wiss-TAY-ree-uh). See *Wisteria*.
Wisteria (wiss-TEE-ree-uh; wiss-TEER-ee-uh as common name). Twining woody vines with colorful flowers.
Woodsia (WOOD-zee-uh). Hardy rockery ferns.
Woodwardia (wood-WAHR-dee-uh). Chain fern.
Wulfenia (wool-FEE-nee-uh). Hardy rockery plants, from proper name.

X

Xanthinus (zan-THY-nus). Yellow.
Xanthisma (zan-THIZ-muh; zan-THISS-muh). Southern prairie plants.
Xanthocarpus (zan-tho-KAHR-pus). Yellow-fruited.
Xanthoceras (zan-THOSS-ur-ass). Hardy Chinese shrub.
Xanthophyllus (zan-tho-FILL-us). Yellow-leaved.
Xanthorrhiza (zan-tho-RY-zuh). Yellow-rooted, zanthorhiza.
Xanthosoma (zan-tho-SO-muh). Greenhouse foliage plants.
Xeranthemum (zee-RAN-theh-mum). Everlasting, immortelle.
Xerophyllum (zee-ro-FIL-um). Turkey beard, elk grass.
Xiphium (ZIF-ee-um; ZY-fee-um). A type of iris.
Xylosma (zy-LOSS-muh). Wood-scented.
Xylosteum (zih-LOSS-tee-um). Hard wood.

Y

Yedoensis (yed-o-EN-siss). From Yeddo, Japan.
Yucca (YUK-uh). Adam's-needle, Spanish dagger, Joshua tree.
Yunnanensis (yoo-na-NEN-siss). From Yunnan, China.

Z

Zamia (ZAY-mee-uh). Coontie.
Zantedeschia (zan-tee-DESS-kee-uh). Calla of florists.
Zanthorhiza (zan-tho-RY-zuh). Yellowroot.
Zanthoxylum (zan-THOK-sih-lum). Prickly ash, Hercules'-club.
Zanzibariensis (zan-zih-bahr-ee-EN-siss). Of Zanzibar, Africa.
Zea (ZEE-uh). Corn, maize.
Zebrina (zee-BRY-nuh). Wandering Jew.
Zebrinus (zee-BRY-nus). Zebra-striped.
Zelkova (zel-KO-vuh). Hardy elm-like tree.
Zenobia (zen-O-bee-uh). White-flowered shrub for forcing.
Zephyranthes (zef-ih-RAN-theez). Zephyr, atamasco and fairy lilies.
Zeylanicus (zee-LAN-ih-kus). Of Ceylon.
Zingiber (ZIN-jih-bur). Ginger.
Zinnia (ZIN-ee-uh). Popular, showy garden annuals.
Zizania (zy-ZAY-nee-uh). Wild rice.
Zizyphus (ZIZ-ih-fus). Jujube.
Zoysia (zo-ISS-ee-uh; ZOY-see-uh). Korean lawn grass.
Zygocactus (zy-go-KAK-tus). Christmas or crab cactus.
Zygopetalum (zy-go-PET-uh-lum). Showy tropical orchids.